Lecture Notes in Physics

Founding Editors: W. Beiglböck, J. Ehlers, K. Hepp, H. Weidenmüller

The Lecture Notes in Physics

The series Lecture Notes in Physics (LNP), founded in 1969, reports new developments in physics research and teaching – quickly and informally, but with a high quality and the explicit aim to summarize and communicate current knowledge in an accessible way. Books published in this series are conceived as bridging material between advanced graduate textbooks and the forefront of research and to serve three purposes:

- to be a compact and modern up-to-date source of reference on a well-defined topic

- to serve as an accessible introduction to the field to postgraduate students and nonspecialist researchers from related areas

- to be a source of advanced teaching material for specialized seminars, courses and schools

Both monographs and multi-author volumes will be considered for publication. Edited volumes should, however, consist of a very limited number of contributions only. Proceedings will not be considered for LNP.

Volumes published in LNP are disseminated both in print and in electronic formats, the electronic archive being available at springerlink.com. The series content is indexed, abstracted and referenced by many abstracting and information services, bibliographic networks, subscription agencies, library networks, and consortia.

Proposals should be sent to a member of the Editorial Board, or directly to the managing editor at Springer:

Christian Caron
Springer Heidelberg
Physics Editorial Department I
Tiergartenstrasse 17
69121 Heidelberg / Germany
christian.caron@springer.com

Richard C. Powell

Symmetry, Group Theory, and the Physical Properties of Crystals

 Springer

Richard C. Powell
Professor Emeritus
University of Arizona
Tucson, AZ
USA
rcpowell@u.arizona.edu

ISSN 0075-8450   e-ISSN 1616-6361
ISBN 978-1-4419-7597-3   e-ISBN 978-1-4419-7598-0
DOI 10.1007/978-1-4419-7598-0
Springer New York Dordrecht Heidelberg London

Printed on acid-free paper

Springer is part of Springer Science+Business Media (www.springer.com)

Preface

Why do we look at some things and think they are beautiful while other things do not appear esthetically pleasing to us? This is a question that has always interested mankind. One answer is given by the following quotation from an early president of the College of New Jersey (now Princeton University):

> "Beauty is found in immaterial things like proportion or uniformity. . . .
> called by various names of regularity, order, uniformity, symmetry,
> proportion, harmony, etc.". . . Jonathan Edwards[1]

Symmetry not only provides the natural harmony that makes something appear beautiful to us, but also is of great value to science because it dictates the physical traits of many objects. Nature itself seems to love beauty since atoms tend to self-assemble into shapes with specific symmetry and crystals grow in geometric lattices. In many cases, if we know the symmetry of something we can predict some of its important properties without having to resort to experimentation or complicated calculations.

One area where the concept of symmetry plays an important role is that of crystalline solids. Crystals, by their very nature, exhibit specific symmetries. Crystalline materials have many important applications in devices based on their electronic, optical, thermal, magnetic, and mechanical properties. Solid state physicists and chemists, as well as material scientists and engineers, have developed rigorous quantum theoretical models to describe these properties and sophisticated measurement techniques to verify these models.

Many times, however, in screening materials for a new application it is useful to be able to quickly and easily determine if a specific material will have the appropriate properties without making detailed calculations or experiments. This can be done by analyzing the symmetry properties of the material. The mathematical formalism that has been developed to accomplish this is called group theory. The symmetry properties of a crystal can be described by a group of mathematical

---

[1] J. Edwards, *Works of Jonathan Edwards* (Banner of Truth Trust, Edinburgh, 1979)

operations. Then using simple group theory procedures, the physical properties of the crystal can be determined.

During the 45 years I have been involved in teaching and research in various areas of solid state physics, I have made extensive use of the concepts of group theory. Yet I have been surprised at how little emphasis this topic receives in any formal educational curriculum. Generally, a student studying solid state physics or chemistry will be exposed to crystal structures early in the semester and then have no further exposure to crystal symmetry until some special topic such as nonlinear optics is discussed. This book focuses on the symmetry of crystals and the description of this symmetry through the use of group theory. Although specific examples are provided of using this formalism to determine both the microscopic and macroscopic properties of materials, the emphasis is on the comprehensive, pervasive nature of symmetry in all areas of solid state science.

The intent of the book is to be a reference source for those doing research or teaching in solid state science and engineering, or a text for a specialty course in group theory applied to the properties of crystals.

Tucson, AZ   Richard C. Powell
June 2010

Contents

Chapter 1
Symmetry in Solids

The intent of this book is to demonstrate the importance of symmetry in determining the properties of solids and the power of using group theory and tensor algebra to elucidate these properties. It is not meant to be a comprehensive text on solid state physics, so many important aspects of condensed matter physics not related to symmetry are not covered here. The book begins by discussing the concepts of symmetry relevant to crystal structures. This is followed by a summary of the basics of group theory and how it is applied to quantum mechanics. Next is a discussion of the description of the macroscopic properties of crystals by tensors and how symmetry determines the form of these tensors. The basic concepts covered in these early chapters are then applied to a series of different examples. There is a discussion of the use of point symmetry in the crystal field theory treatment of point defects in solids. Next is a discussion of crystal symmetry in determining the optical properties of solids, followed by a chapter on the nonlinear optical properties of solids. Then the role of symmetry in treating lattice vibrations is described. The last chapter discusses the effects of translational symmetry on electronic energy bands in solids. The emphasis throughout the book shows how group theory and tensor algebra can provide important information about the properties of a system without resorting to first principal calculations.

1.1 Symmetry

The word "symmetry" commonly refers to the fact that the shapes and dimensions of some objects repeat themselves in different parts of the object or when the object is viewed from different perspectives. Symmetry pervades every aspect of our lives (Fig. 1.1). In the realm of art and architecture, symmetry gives the object a certain esthetically pleasing quality. Many musical compositions of classical composers such as J.S. Bach show symmetry in their structure by repeating the same theme many times throughout the piece, sometimes with variations. In the realm of science, symmetry determines some of the fundamental physical and chemical properties of an object.

R.C. Powell, *Symmetry, Group Theory, and the Physical Properties of Crystals*,
Lecture Notes in Physics 824, DOI 10.1007/978-1-4419-7598-0_1,
© Springer Science+Business Media, LLC 2010

Fig. 1.1 Navajo weavers are
famous for being able to
produce patterns that are
symmetric about horizontal
and vertical center lines
without having any predrawn
plan to follow

The concept of symmetry in science is important in theories and models as well as the shape of discrete objects. Nature likes to have things symmetric. The symmetry of nature plays a critical role in everything from our understanding of the nature of elementary particles to our models of the structure of galaxies in the universe. Almost all of the laws of nature have their root in some type of symmetry. Because of this, if we elucidate the symmetry of a physical system we can predict many of its physical properties. Nowhere is symmetry more important than in understanding the physical and chemical properties of solids.

If a change is made to a physical system (either a discrete object or a mathematical formula describing a physical property), this is called a *transformation*. If a system appears to be exactly the same before and after the transformation, it is said to be *invariant* under that transformation. The symmetry of the system is made up of all of the transformation operations that leave the system invariant. This can be applied to both classical and quantum physics and is important in understanding both the atomic scale and the macroscopic properties of solids. The laws of physics relevant to a system must remain invariant under the symmetry transformations for the system. For example, the Hamiltonian operator describing the total energy of a quantum mechanical system must be invariant under any symmetry operation of the system.

A spatial symmetry transformation acts about a *symmetry element*. A symmetry element can be a point, an axis, or a plane of symmetry resulting in inversion, rotation,

Fig. 1.2 Symmetry elements
of a square array of four
equivalent atoms

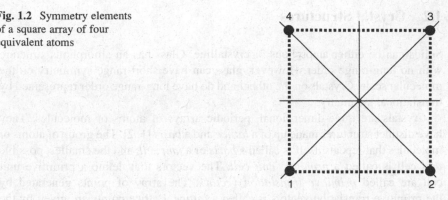

and mirror types of transformations. As an example, consider a two-dimensional array of four equivalent atoms arranged at the corners of a square as shown in Fig. 1.2. The point at the center of the square is a symmetry element for an inversion operation. It takes the atom at point 1 into point 3, the atom at point 3 into point 1, the atom at point 2 into point 4, and the atom at point 4 into point 2. An axis at this center point perpendicular to the plane of the paper is a symmetry element for rotation operations. Rotations of the square about this axis by 90°, 180°, 270°, and 360° all leave the arrangement of atoms invariant. For example, a 90° counterclockwise rotation about this axis takes the atom at point 1 to point 2, the atom at point 2 to point 3, the atom at point 3 to point 4, and the atom at point 4 to point 1. Four mirror planes perpendicular to the plane of the paper and containing the symmetry axis are symmetry elements. Two of these bisect the sides of the square while the other two go through opposite corners. As an example, the mirror plane going from point 1 through the center to point 3 will leave the atoms at points 1 and 3 invariant while interchanging the atoms at points 2 and 4. In three dimensions it is also possible to have a combined symmetry elements of rotation about an axis followed by reflection in a plane perpendicular to that axis.

Mathematically these operations can be represented by matrices operating on the coordinates used to describe the physical system of interest. The physical properties of matter can be described by tensors of specific ranks. An nth rank tensor in three-dimensional space is a mathematical entity with n indices and 3^n components that obey specific transformation rules. A zero-rank tensor has no indices and is referred to as a scalar. A first-rank tensor has one index and three components and is called a vector. A second-rank tensor has two indices and 9 components and is called a matrix. General tensors are extensions of this progression to higher orders. The mathematical fields of *group theory* and *tensor algebra* have been developed to describe the symmetry properties of a system. Group theory is a powerful tool in physics. It allows the determination of many of the physical properties of a system without resorting to rigorous first principal calculations of these properties. However, group theory provides only qualitative information about whether or not a system possesses a particular property; it can not predict the magnitude of the property.

1.2 Crystal Structures

Solids can be either amorphous or crystalline. Glass has an amorphous structure
with no long-range order. However, glass can have short-range symmetry on the
molecular scale. Crystals on the other hand do have long-range order represented by
translational symmetry.

Crystals are three-dimensional, periodic arrays of atoms or molecules. They
have distinct structures made up of a *lattice* and a *basis* [1, 2]. The group of atoms or
molecules that repeats itself is called a *basis* or a *unit cell*, and the smallest possible
unit cell is called a *primitive unit cell*. The vectors that define a primitive unit
cell are called *primitive translation vectors*. The array of points generated by
the primitive translation vectors is called a *lattice*. Lattice points are given by the
equation

$$\mathbf{T}_n = n_1\mathbf{a} + n_2\mathbf{b} + n_3\mathbf{c}, \tag{1.1}$$

where $\mathbf{a}$, $\mathbf{b}$, and $\mathbf{c}$ are the primitive translation vectors and the n_i are integers.
The arrangement of atoms or molecules will look exactly the same from any lattice
point.

Each atom that is a part of the basis is associated with a specific lattice point but
all atoms do not appear on a lattice point. A simple example of this is shown in
Fig. 1.3 for a basis of atoms A and B on a two-dimensional square lattice. The A
atoms are located on each lattice point designated $\vec{T}_A = 0$ while the B atoms are
between lattice points at positions designated $\vec{T}_B = (1/2)(\mathbf{a} + \mathbf{b})$.

Any operation performed on a crystal that carries the crystal structure into itself
is part of the symmetry group for that crystal. This may include translations,

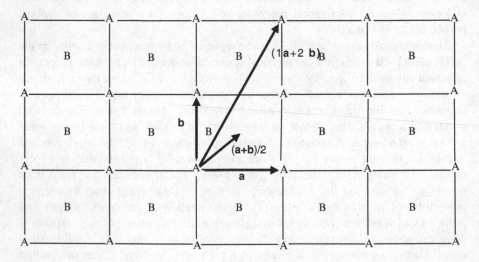

Fig. 1.3 Square lattice with basis atoms A and B

reflections through planes, rotations about axes, inversion through a point, and combinations of these operations. The fundamental types of crystal lattices are defined by their symmetry operations. These include *translation group*, *point group*, and *space group* symmetries. The translation group has operations given by $\{E|\vec{T}\}$ where E is the identity rotation operation and $\vec{T}$ is a translation operation that leaves the crystal invariant. Examples of translation operations shown in Fig. 1.3 are the primitive lattice vectors $\vec{T} = 1\vec{a}, \vec{T} = 1\vec{b}$, and $\vec{T} = 1\vec{a} + 2\vec{b}$. The point group has operations given by $\{\alpha|0\}$ where α is a symmetry operation at a point that leaves the crystal invariant with no translation. A lattice point group for a crystal can have a two-, three-, four-, or sixfold axis of rotation plus reflections and inversion operations.

For the example shown in Fig. 1.3, the point group symmetry operations at lattice point A are those shown in Fig. 1.2 and discussed earlier. The space group has operations $\{\alpha|\vec{T}\}$ that leave the crystal invariant. If all operations $\{\alpha|0\}$ in a space group form a subgroup the space group is called *symmorphic*. In the examples given above, the point group operations for the array of equivalent atoms in Fig. 1.2 combine with the translation operations of the lattice shown in Fig. 1.3 to form a symmorphic space group. If the atoms on the lattice are not all equivalent as shown in Fig. 1.4, the space group may not be symmorphic and some symmetry operations involve the combination of a translation with a point group operation. For example in Fig. 1.4a, the 90°, 180°, and 270° rotations are still symmetry operations but the four mirror planes only leave the array invariant if they are combined with the translation operation $\vec{T}_B = 1/2(\mathbf{a} + \mathbf{b})$. These combined translation–reflection operations $\{\sigma_i|\vec{T}_B\}$, where σ_i represents one of the four mirror planes, are referred to as *glide planes*. Similarly, for the example array of atoms shown in Fig. 1.4b, the rotations of 90° and 270° only leave the system invariant when combined with the translation operation $\vec{T}_B = (1/2)(\mathbf{a} + \mathbf{b})$. These combined translation–rotation operations $\{C_i|\vec{T}_B\}$, where C_i represents one of the two rotation operations, are referred to as *screw axes*. There is an important restriction on glide plane and screw axis symmetry operations in three dimensions that does not apply to the two-dimensional examples discussed above [3]. In three dimensions the translation

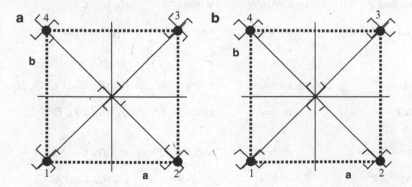

Fig. 1.4 A lattice with a basis of nonequivalent shapes

part of the combined operation must be in the mirror plane of the glide operation or parallel to the rotation axis for the screw operation. If the array of atoms in Fig. 1.4a continues out of the page making it three dimensional, then it could be expressed as $\{\sigma_{13}|(\vec{a}+\vec{b}+\vec{c})/2\}$ where the vector $(\vec{a}+\vec{b}+\vec{c})/2$ lies in the σ_{13} plane. For the situation shown in Fig. 1.4b, a rotation of $180°$ about the 1–3 axis leaves the array invariant only if it is combined with a translation of $\vec{T}_B = (1/2)(\mathbf{a}+\mathbf{b})$, which is parallel to the 1–3 axis and therefore can be a screw axis in three dimensions.

There are 14 different types of crystal lattices found in nature, referred to as *Bravais lattices* [1, 2, 4]. These are defined by the primitive translation vectors and the angles between them where α is the angle between **b** and **c**, β is the angle between **c** and **a**, and γ is the angle between **a** and **b**. The magnitudes of the translation vectors are called the *lattice parameters*. The conventional cells (not necessarily primitive) are shown in Fig. 1.5 for each type of lattice organized into seven *crystal systems*. Each crystal system has a distinctly different shape. The relationships between the lattice parameters and the angles for each of these are given in Table 1.1. The Bravais unit cell is the smallest unit cell that exhibits the symmetry of the structure.

The *primitive Bravais lattices* have one lattice site at position (0,0,0). There are three types of *nonprimitive Bravais lattices*. The two-face-centered lattice designated by C has lattice sites at positions (0,0,0,) and (1/2,1/2,0). The internally centered lattice designated by I has lattice sites at positions (0,0,0) and (1/2,1/2,1/2). The all-face-centered lattice designated by F has lattice sites at positions (0,0,0), (1/2,1/2,0), (1/2,0,1/2), and (0,1/2,1/2). There are seven types of primitive Bravais lattices and seven types of nonprimitive lattices that make up the fourteen crystal symmetries. In dealing with space groups discussed below it is important to further divide the two-face-centered lattice structure into three types

Table 1.1 Three-dimensional crystal lattices

Crystal systems and Bravais lattices	Lattice parameters	Angles	Point group symmetry (Crystal classes)
Triclinic P	$a\neq b\neq c$	$\alpha\neq\beta\neq\gamma\neq90°$	C_i (C_1)
Monoclinic P	$a\neq b\neq c$	$\alpha=\gamma=90°\neq\beta$	C_{2h} (C_2, C_s)
C			
Orthorhombic P	$a\neq b\neq c$	$\alpha=\beta=\gamma=90°$	D_{2h} (D_2, C_{2v})
C			
I			
F			
Tetragonal P	$a=b\neq c$	$\alpha=\beta=\gamma=90°$	D_{4h} (D_4, C_{4v}, C_{4h}, C_4, D_{2d}, S_4)
I			
Cubic P	$a=b=c$	$\alpha=\beta=\gamma=90°$	O_h (O, T_d, T_h, T)
I			
F			
Trigonal R	$a=b=c$	$\alpha=\beta=\gamma<120°$, $\neq90°$	D_{3d} (D_3, C_{3v}, S_6, C_3)
Hexagonal P	$a=b\neq c$	$\alpha=\beta=90°$, $\gamma=120°$	D_{6h} (D_6, C_{6v}, C_{6h}, C_6, D_{3h}, C_{3h})

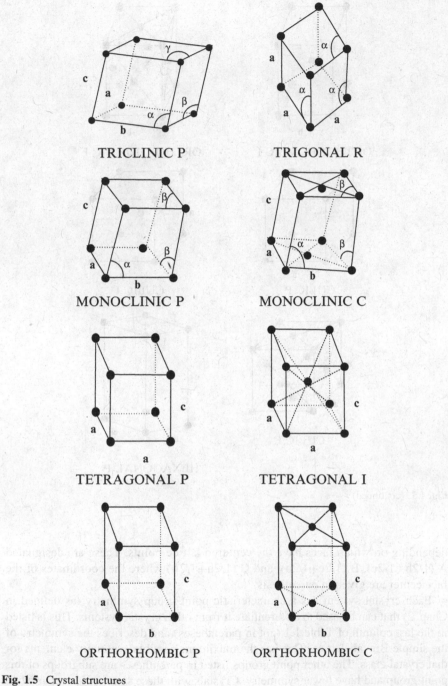

Fig. 1.5 Crystal structures

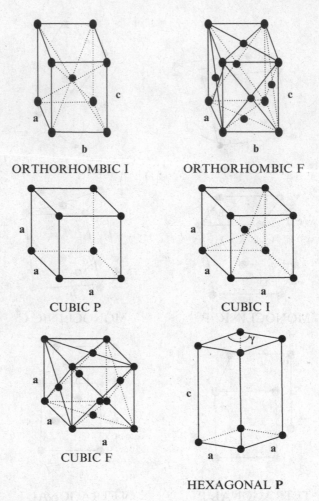

ORTHORHOMBIC I   ORTHORHOMBIC F

CUBIC P   CUBIC I

CUBIC F   HEXAGONAL P

Fig. 1.5 (continued)

depending on which faces have the centered lattice points. These are designated A $(1/2\mathbf{b}+1/2\mathbf{c})$, B $(1/2\mathbf{c}+1/2\mathbf{a})$, and C $(1/2\mathbf{a}+1/2\mathbf{b})$ where the coordinates of the face center are given in parenthesis.

Each crystal system has a characteristic point group symmetry (as defined in Chap. 2) that can be used to differentiate it from other crystal systems. This is listed in the last column of Table 1.1 (not in parentheses) and describes the symmetry of the simple Bravais lattice. This has the maximum possible symmetry elements for that crystal class. The other point groups listed in parentheses are subgroups of this point group and have lower symmetry. Crystals with these symmetries occur when the molecules or atoms placed as a basis on the Bravais lattice have lower symmetry than the lattice itself. Chapter 2 has a detailed discussion of these point group symmetries. There are two types of point groups for the triclinic lattice, three types each for the monoclinic and orthorhombic systems, seven each for the tetragonal and hexagonal systems, and five each for trigonal and cubic systems. This gives a total of 32 point groups associated with the 14 Bravais lattices as summarized in Table 1.1.

There are five types of symmetry axes that occur in crystals representing n-fold rotations. These are one-, two-, three-, four-, and sixfold axes representing $360/ns>$degrees of rotation designated as C_n. In addition there are mirror planes perpendicular to the major rotation axis designated as σ_h and mirror planes containing the major rotation axis designated as σ_v. Mirror planes diagonal to the rotation axes are designated as σ_d. The identity and inversion operations are designated E and i. Combined rotation/reflection operations with the mirror plane perpendicular to the rotation axis are designated S_n. The point groups in Table 1.1 are given in the Schoenflies notation [5]. Groups with nth order cyclic rotations about a single axis are designated C_n. Adding a mirror reflection element normal to the rotation axis gives groups designated C_{nh}. C_{nv} designates groups with mirror planes containing the rotation axis. Adding a twofold rotation element perpendicular to the major rotation axis gives groups designated as D_n. D_{nd} groups have additional vertical symmetry planes between the twofold axes. In addition, there are tetrahedral groups T, T_d, and T_h as well as octahedral groups O and O_h. The resulting lattice systems are shown in Fig. 1.5.

The *triclinic* system is identified by the fact that it has no rotation or reflection symmetry elements. It is the least symmetric of all the lattice systems. It can be characterized as a parallelepiped with unequal edges and unequal angles.

One twofold axis in the b-direction identifies the *monoclinic* system. In addition it has a mirror plane perpendicular to this axis. The three edges of the unit cell are unequal in length with two being perpendicular to the symmetry axis. This may occur as a simple primitive lattice or a base-center cell with points at the center of the faces parallel to the reflection plane.

Three mutually orthogonal twofold axes identify an *orthorhombic* system. In addition there are reflection planes perpendicular to these axes. The three edges of the unit cell are unequal in length but are all mutually orthogonal. This gives rise to four Bravais lattices: primitive, base centered, body centered, and face centered.

One fourfold rotation axis is the characteristic symmetry for the *tetragonal* system. This is in addition to the twofold axes and mirror planes found in the orthorhombic system. This comes about because two of the edges of the lattice cell in this system are equal. This leads to a primitive lattice and a body-centered lattice.

One threefold rotation axes is the characteristic symmetry for the *trigonal* system. The shape of the lattice cell is a rhombohedron with equal sides and equal angles (none of which are $90°$). There are three twofold rotation axes perpendicular to the trigonal axis, the inversion operation, three reflection planes containing the trigonal axis, and two combined rotations of $60°$ about the trigonal axis and reflection in a plane perpendicular to this axis. There is a second type of lattice in the trigonal system designated P that is equivalent to the hexagonal lattice so it is not shown as a separate lattice in Fig. 1.5.

The characteristic symmetry for the *hexagonal* system is one sixfold axes. It also has threefold and twofold rotations about this axis and twofold rotation axes perpendicular to this axis. There is also inversion symmetry. Two of the edges of the lattice cell have equal lengths, two of the angles are 90°, and the third one is 120°. This has only the primitive lattice structure.

The *cubic* system is identified by the presence of either four equivalent threefold axes or three equivalent fourfold axes. In addition it contains several twofold rotation axes, mirror planes, and inversion symmetry. These can be found in simple cubic, body-centered-cubic, and face-centered-cubic lattices. This is the most symmetric type of crystal structure.

Twenty-seven of the 32 point group symmetries have a preferred axis of symmetry. Because of this, it is useful to represent each of the these noncubic point group symmetries by *stereograms* that specifically show the symmetry elements of the group [6]. These are circles with the z-axis coming out of the page, ● representing a point on top of the page, ○ representing a point below the plane of the page, and ◐ representing points above and below the plane of the page. These are useful in describing an inversion operation. A mirror plane is a full line. Rotation axes are represented by ▮, ▲, ◆, and ⬣ for twofold, threefold, fourfold, and sixfold rotations, respectively. These 27 sterograms are shown in Fig. 1.6.

The five cubic point symmetry groups have equivalent, orthogonal axes of symmetry instead of a single, preferred axis of symmetry. Thus it is not useful to try to represent the symmetry elements of these groups by simple two-dimensional stereograms such as those shown in Fig. 1.6. Instead these must be visualized using a three-dimensional cube. Figure 1.7 shows the symmetry elements of the O and O_h cubic groups. The notations for the rotation axes and the inversion operation are the same as those used in Fig. 1.6. There are three twofold and fourfold axes of symmetry parallel to the cube edges. There are also six twofold axes parallel to the face diagonals (only two are shown in the figure for simplicity). There are four threefold axes of symmetry about the body diagonals (only two are shown in the figure for simplicity). Inversion symmetry is present in O_h but not in O and is represented by the open and filled circles as in Fig. 1.7. There are three reflection planes perpendicular to the cube edges. There are six reflection planes parallel to the face diagonals (only one of which is shown for simplicity). In addition there are combined operations consisting of a ±90° rotation about the axis parallel to the cube edges followed by reflection through a plane perpendicular to the rotation axis. This is represented in the figure by ◆. Finally, there are combined rotations of ±120° about the body diagonals followed by reflections through the planes perpendicular to these axes. These are represented in the figure by ▲. Only two of these are shown in the figure for simplicity.

The final three groups of the cubic class can be visualized by carving out a tetrahedron from the cube shown in Fig. 1.7. This is shown in Fig. 1.8. Then comparing the symmetry operations shown in Fig. 1.8 with Fig. 1.9 shows that the tetrahedron still has the three twofold symmetry axes parallel to the face edges and the eight threefold symmetry axes about the body diagonals. The T point group is made up of the identity plus these pure rotations as shown in Fig. 1.9. The T_h group contains the elements of T

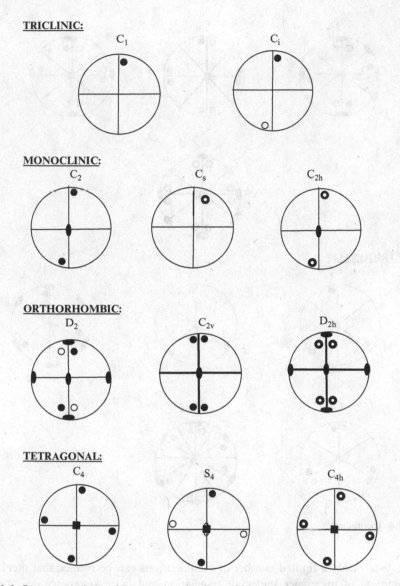

TRICLINIC:
C_1 C_i

MONOCLINIC:
C_2 C_s C_{2h}

ORTHORHOMBIC:
D_2 C_{2v} D_{2h}

TETRAGONAL:
C_4 S_4 C_{4h}

Fig. 1.6 Stereograms for the 27 noncubic crystallographic point groups

plus the inversion operation and its product with each of the other elements, as shown Fig. 1.9. The T_d group has the elements of T plus a mirror plane containing one edge of the tetrahedron and bisecting the opposite edge and the elements obtained from multiplying this mirror element with each of the other elements of T. This is also shown in Fig. 1.9.

One of the 14 Bravais lattices having one of the seven symmetry types coupled with one of the 32 point groups can be used to describe the total symmetry of

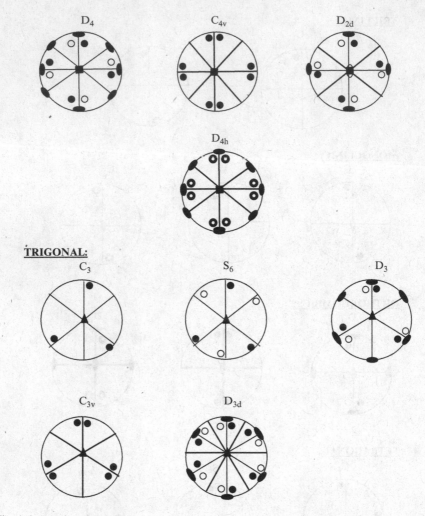

Fig. 1.6 (continued)

any crystal. Only a limited number of combinations can be formed that meet the requirement of invariance under all translation and point symmetry operations. These are called *space groups* [3, 7]. The symmetry operations of a specific space group include the point group operations and the translation operations, and any combined translational rotation or reflection operations. These can be used to obtain 66 of the 73 symmorphic space groups. The other seven symmorphic space groups have point groups that have two possible nonequivalent orientations on the Bravais lattice. For example C_{2v} can be on an A face centered lattice with operations such as glide planes and screw axes. As discussed above, the translations for screw and glide operations are nonprimitive lattice vectors, and in three dimensions their direction is parallel to the screw rotation axis or in the mirror plane of the glide

HEXAGONAL:

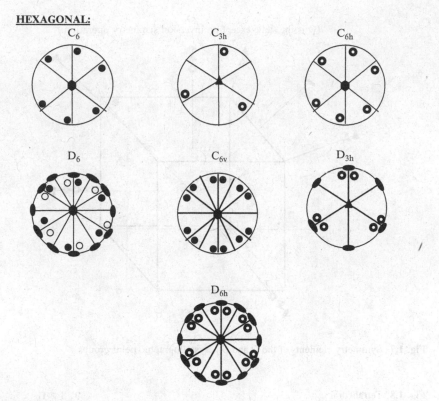

Fig. 1.6 (continued)

operation. As mentioned previously, space groups that do not contain these types of operations are referred to as symmorphic while space groups containing screw or glide operations are called asymmorphic.

There are 230 possible space groups of which 73 are symmorphic and 157 are asymmorphic. Symmorphic space groups are made up of all the symmetry operations of a crystallographic point group $\{\alpha|0\}$ combined with all of the translational symmetry operations of a Bravais lattice $\{E|\vec{T}\}$ to give a complete set of symmetry operations for the crystal $\{\alpha|\vec{T}\}$. This simple method combines point groups and centered lattice or a C-centered lattice. The latter is centered along a twofold rotation axis while the former is not. This results in two nonequivalent space groups. Similarly, the point groups D_{2d}, D_{3h}, D_3, C_{3v}, and D_{3d} can each generate more than one space group because of different types of orientations on their Bravais lattices. For nonsymmorphic space groups, some of the point group operations are combined with translation operations to give operators of the form $\{\alpha|\vec{G}(\alpha)\}$, where $\vec{G}(\alpha)$ are not lattice translation vectors. There are two space groups in the triclinic system, 13 in the monoclinic system, 59 in the orthorhombic system, 68 in the tetragonal system, 25 in the trigonal system, 27 in the hexagonal system, and 36 in the cubic system.

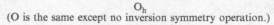

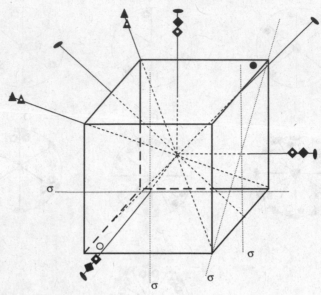

Fig. 1.7 Symmetry elements of the O and O_h crystallographic point groups

Fig. 1.8 Tetrahedral
symmetry as part of a cube

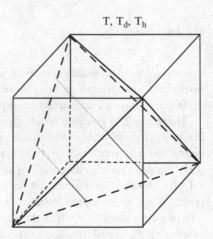

The 230 space groups are listed in Table 1.2. There are two types of notations commonly used to designate space groups of crystals. The first of these follows the Schoenflies notation for the point groups with superscripts used to distinguish among space groups combined with different types of lattices and involving screw or glide operations. This notation explicitly designates a specific crystal class which

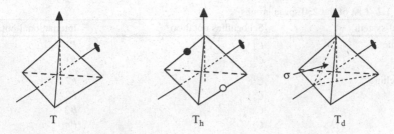

Fig. 1.9 Tetrahedral symmetry groups T, T_d, T_h

implies one of the crystal systems listed in Table 1.1. However, the superscripts are not helpful in identifying a specific Bravais lattice or the presence of any combined rotation/translation symmetry elements. For example, the space groups O_h^i where $i = 1,\ldots,4$ are associated with a simple cubic lattice, the space groups designated O_h^i where $i = 5,\ldots,8$ are associated with a face-centered cubic lattice, and O_h^i where $i = 9$ and 10 are associated with a body-centered cubic lattice.

The second type of notation is called the international notation. A specific space group designation begins with a capital letter that designates the type of lattice centering shown in Fig. 1.5. For face-centered lattices the letters A, B, and C designate the specific face where the centering occurs as opposed to the generic designation C used in Fig. 1.5. Also the letter R occurs in the trigonal crystal system to designate rhombohedral lattice while the trigonal lattice designated by the letter P is essentially equivalent to a hexagonal P lattice. The next part of the space group designation is the point group symbol for the crystal class. In this notation, the numbers designate the primary rotation axes (one-, two- three-, four-, or six-fold), letter m designates a mirror plane containing the rotation axis, and /m a mirror plane perpendicular to the rotation axis. A bar over an axis number designates a rotation–reflection combined operation. A subscript on a rotation axis indicates that it is a screw operation. For example, 4_1, 4_2, and 4_3 designate a fourfold screw axis with translations of ¼, ½, and ¾ of a lattice vector. (Note that the translation part of a screw operation is always a submultiple of the rotation part.) The letters a, b, c, n, and d designate glade plane operations involving translations of one half a lattice translation in a specific direction before reflection. Knowing the point group and type of centering gives the specific Bravais lattice for the space group. As an example, the space group O_h^5 in Schoenflies notation is $Fm3m$ in the international notation. This shows the space group to be in the face-centered cubic crystal system in the $m3m$ crystal class with no screw axes for glide plane symmetry operations. The details of this type of designation are given in [5]. Both the Schoenflies and international notations are given in Table 1.2.

1.3 Symmetry in Reciprocal Space

Quasiparticles on a periodic crystal lattice (such as electrons or phonons) are described by eigenfunctions of the form

Table 1.2 List of the 230 space groups

Crystal system	Schoenflies notation	International notation
Triclinic	C_1^1	$P1$
	C_i^1	$P\bar{1}$
Monoclinic	C_2^1	$P2$
	C_2^2	$P2_1$
	C_2^3	$B2$
	C_s^1	Pm
	C_s^2	Pb
	C_s^3	Bm
	C_s^4	Bb
	C_{2h}^1	$P2/m$
	C_{2h}^2	$P2_1/m$
	C_{2h}^3	$B2/m$
	C_{2h}^4	$P2/b$
	C_{2h}^5	$P2_1/b$
	C_{2h}^6	$B2/b$
Orthorhombic	D_2^1	$P222$
	D_2^2	$P222_1$
	D_2^3	$P2_12_12$
	D_2^4	$P2_12_12_1$
	D_2^5	$C222_1$
	D_2^6	$C222$
	D_2^7	$F222$
	D_2^8	$I222$
	D_2^9	$I2_12_12_1$
	C_{2V}^1	$Pmm2$
	C_{2V}^2	$Pmc2_1$
	C_{2V}^3	$Pcc2$
	C_{2V}^4	$Pma2$
	C_{2V}^5	$Pca2_1$
	C_{2V}^6	$Pnc2$
	C_{2V}^7	$Pmn2_1$
	C_{2V}^8	$Pba2$
	C_{2V}^9	$Pna2_1$
	C_{2V}^{10}	$Pnn2$
	C_{2V}^{11}	$Cmm2$
	C_{2V}^{12}	$Cmc2_1$
	C_{2V}^{13}	$Ccc2$
	C_{2V}^{14}	$Amm2$
	C_{2V}^{15}	$Abm2$
	C_{2V}^{16}	$Ama2$
	C_{2V}^{17}	$Aba2$
	C_{2V}^{18}	$Fmm2$
	C_{2V}^{19}	$Fdd2$
	C_{2V}^{20}	$Imm2$
	C_{2V}^{21}	$Iba2$
	C_{2V}^{22}	$Ima2$

(continued)

Table 1.2 (continued)

Crystal system	Schoenflies notation	International notation
	D_{2h}^1	Pmmm
	D_{2h}^2	Pnnn
	D_{2h}^3	Pccm
	D_{2h}^4	Pban
	D_{2h}^5	Pmma
	D_{2h}^6	Pnna
	D_{2h}^7	Pmna
	D_{2h}^8	Pcca
	D_{2h}^9	Pbam
	D_{2h}^{10}	Pccn
	D_{2h}^{11}	Pbcm
	D_{2h}^{12}	Pnnm
	D_{2h}^{13}	Pmmn
	D_{2h}^{14}	Pbcn
	D_{2h}^{15}	Pbca
	D_{2h}^{16}	Pnma
	D_{2h}^{17}	Cmcm
	D_{2h}^{18}	Cmca
	D_{2h}^{19}	Cmmm
	D_{2h}^{20}	Cccm
	D_{2h}^{21}	Cmma
	D_{2h}^{22}	Ccca
	D_{2h}^{23}	Fmmm
	D_{2h}^{24}	Fddd
	D_{2h}^{25}	Immm
	D_{2h}^{26}	Ibam
	D_{2h}^{27}	Ibca
	D_{2h}^{28}	Imma
Tetragonal	C_4^1	P4
	C_4^2	$P4_1$
	C_4^3	$P4_2$
	C_4^4	$P4_3$
	C_4^5	I4
	C_4^6	$I4_1$
	S_4^1	$P\overline{4}$
	S_4^2	$I\overline{4}$
	C_{4h}^1	P4/m
	C_{4h}^2	$P4_2/m$
	C_{4h}^3	P4/n
	C_{4h}^4	$P4_2/n$
	C_{4h}^5	I4/m
	C_{4h}^6	$P4_1/a$
	D_4^1	P422
	D_4^2	$P42_12$
	D_4^3	$P4_122$
	D_4^4	$P4_12_12$

(continued)

Table 1.2 (continued)

Crystal system	Schoenflies notation	International notation
	D_4^5	$P4_222$
	D_4^6	$P4_22_12$
	D_4^7	$P4_322$
	D_4^8	$P4_32_12$
	D_4^9	$I422$
	D_4^{10}	$I4_122$
	C_{4v}^1	$P4mm$
	C_{4v}^2	$P4bm$
	C_{4v}^3	$P4_2cm$
	C_{4v}^4	$P4_2nm$
	C_{4v}^5	$P4cc$
	C_{4v}^6	$P4nc$
	C_{4v}^7	$P4_2mc$
	C_{4v}^8	$P4_2bc$
	C_{4v}^9	$I4mm$
	C_{4v}^{10}	$I4cm$
	C_{4v}^{11}	$I4_1md$
	C_{4v}^{12}	$I4_1cd$
	D_{2d}^1	$P\bar{4}2m$
	D_{2d}^2	$P\bar{4}2c$
	D_{2d}^3	$P\bar{4}2_1m$
	D_{2d}^4	$P\bar{4}2_1c$
	D_{2d}^5	$P\bar{4}m2$
	D_{2d}^6	$P\bar{4}c2$
	D_{2d}^7	$P\bar{4}b2$
	D_{2d}^8	$P\bar{4}n2$
	D_{2d}^9	$I\bar{4}m2$
	D_{2d}^{10}	$I\bar{4}c2$
	D_{2d}^{11}	$I\bar{4}2m$
	D_{2d}^{12}	$I\bar{4}2d$
	D_{4h}^1	$P4/mmm$
	D_{4h}^2	$P4/mcc$
	D_{4h}^3	$P4/nbm$
	D_{4h}^4	$P4/nnc$
	D_{4h}^5	$P4/mbm$
	D_{4h}^6	$P4/mnc$
	D_{4h}^7	$P4/nmm$
	D_{4h}^8	$P4/ncc$
	D_{4h}^9	$P4_2/mmc$
	D_{4h}^{10}	$P4_2/mcm$
	D_{4h}^{11}	$P4_2/nbc$
	D_{4h}^{12}	$P4_2/nnm$
	D_{4h}^{13}	$P4_2/mbc$
	D_{4h}^{14}	$P4_2/mnm$
	D_{4h}^{15}	$P4_2/nmc$
	D_{4h}^{16}	$P4_2/ncm$

(continued)

Table 1.2 (continued)

Crystal system	Schoenflies notation	International notation
	D_{4h}^{17}	$I4/mmm$
	D_{4h}^{18}	$I4/mcm$
	D_{4h}^{19}	$I4_1/amd$
	D_{4h}^{20}	$I4_1/acd$
Trigonal	C_3^1	$P3$
	C_3^2	$P3_1$
	C_3^3	$P3_2$
	C_3^4	$R3$
	C_{3i}^1	$P\bar{3}$
	C_{3i}^2	$R\bar{3}$
	D_3^1	$P312$
	D_3^2	$P321$
	D_3^3	$P3_112$
	D_3^4	$P3_121$
	D_3^5	$P3_212$
	D_3^6	$P3_221$
	D_3^7	$R32$
	C_{3v}^1	$P3m1$
	C_{3v}^2	$P31m$
	C_{3v}^3	$P3c1$
	C_{3v}^4	$P31c$
	C_{3v}^5	$R3m$
	C_{3v}^6	$R3c$
	D_{3d}^1	$P\bar{3}1m$
	D_{3d}^2	$P\bar{3}1c$
	D_{3d}^3	$P\bar{3}m1$
	D_{3d}^4	$P\bar{3}c1$
	D_{3d}^5	$R\bar{3}m$
	D_{3d}^6	$P\bar{3}c$
Hexagonal	C_6^1	$P6$
	C_6^2	$P6_1$
	C_6^3	$P6_5$
	C_6^4	$P6_2$
	C_6^5	$P6_4$
	C_6^6	$P6_3$
	C_{3h}^1	$P\bar{6}$
	C_{6h}^1	$P6/m$
	C_{6h}^2	$P6_3m$
	D_6^1	$P622$
	D_6^2	$P6_122$
	D_6^3	$P6_522$
	D_6^4	$P6_222$
	D_6^5	$P6_422$
	D_6^6	$P6_322$
	C_{6v}^1	$P6mm$

(continued)

Table 1.2 (continued)

Crystal system	Schoenflies notation	International notation
	C_{6v}^2	$P6cc$
	C_{6v}^3	$P6_3cm$
	C_{6v}^4	$P6_3mc$
	D_{3h}^1	$P\bar{6}m2$
	D_{3h}^2	$P\bar{6}c2$
	D_{3h}^3	$P\bar{6}2m$
	D_{3h}^4	$P\bar{6}2c$
	D_{6h}^1	$P6/mmm$
	D_{6h}^2	$P6/mcc$
	D_{6h}^3	$P6_3/\dot{m}cm$
	D_{6h}^4	$P6_3/mmc$
Cubic	T^1	$P23$
	T^2	$F23$
	T^3	$I23$
	T^4	$P2_13$
	T^5	$I2_13$
	T_h^1	$Pm3$
	T_h^2	$Pn3$
	T_h^3	$Fm3$
	T_h^4	$Fd3$
	T_h^5	$Im3$
	T_h^6	$Pa3$
	T_h^7	$Ia3$
	T_d^1	$P\bar{4}3m$
	T_d^2	$F\bar{4}3m$
	T_d^3	$I\bar{4}3m$
	T_d^4	$P\bar{4}3n$
	T_d^5	$F\bar{4}3c$
	T_d^6	$I\bar{4}3d$
	O^1	$P432$
	O^2	$P4_232$
	O^3	$F432$
	O^4	$F4_132$
	O^5	$I432$
	O^6	$P4_332$
	O^7	$P4_132$
	O^8	$I4_132$
	O_h^1	$Pm3m$
	O_h^2	$Pn3n$
	O_h^3	$Pm3n$
	O_h^4	$Pn3m$
	O_h^5	$Fm3m$
	O_h^6	$Fm3c$
	O_h^7	$Fd3m$
	O_h^8	$Fd3c$
	O_h^9	$Im3m$
	O_h^{10}	$Ia3d$

$$\psi(\mathbf{r}) = u(\mathbf{r})e^{i\mathbf{k}\cdot\mathbf{r}}, \tag{1.2}$$

where $\mathbf{k}$ is the wave vector with magnitude $2\pi/\lambda$ and λ is the wavelength and $u(\mathbf{r})$ is a function that has the periodicity of the lattice. To work with wave vectors that have the periodicity of the lattice, it is useful to construct a reciprocal lattice [1, 2, 3, 7]. This is done by defining three vectors $\mathbf{b}_1$, $\mathbf{b}_2$, and $\mathbf{b}_3$ with respect to the primitive translation vectors given in (1.1):

$$\mathbf{t}_i \cdot \mathbf{b}j = \delta_{ij}, \tag{1.3}$$

so that

$$\mathbf{b}_1 = \frac{\mathbf{t}_2 \times \mathbf{t}_3}{\mathbf{t}_1 \cdot (\mathbf{t}_2 \times \mathbf{t}_3)}, \quad \mathbf{b}_2 = \frac{\mathbf{t}_3 \times \mathbf{t}_1}{\mathbf{t}_1 \cdot (\mathbf{t}_2 \times \mathbf{t}_3)}, \quad \mathbf{b}_3 = \frac{\mathbf{t}_1 \times \mathbf{t}_2}{\mathbf{t}_1 \cdot (\mathbf{t}_2 \times \mathbf{t}_3)}. \tag{1.4}$$

The volume of the unit cell in reciprocal space is $\mathbf{b}_1 \cdot (\mathbf{b}_2 \times \mathbf{b}_3)$ which is the reciprocal of the volume of the unit cell in ordinary space, $\mathbf{t}_1 \cdot (\mathbf{t}_2 \times \mathbf{t}_3)$. Using this construction, the unit cells in reciprocal space can be formed for each of the Bravais lattices in ordinary space. These are called *Brillouin zones*. These are constructed by drawing the vectors $\mathbf{K}$ defining the reciprocal lattice and then bisecting each of these with planes perpendicular to $\mathbf{K}$. The shape enclosed by these planes is the *first Brillouin zone*. This zone is repeated throughout the reciprocal lattice by translating it by reciprocal lattice vectors. Any point at position $\mathbf{k}$ in a given Brillouin zone is equivalent to the point defined by $\mathbf{k}$ in the first Brillouin zone.

A wave vector that has the correct periodicity in reciprocal space can then be expressed as

$$\mathbf{K}_h = 2\pi(h_1\mathbf{b}_1 + h_2\mathbf{b}_2 + h_3\mathbf{b}_3), \tag{1.5}$$

where the h_i represent integers. The vectors $(h_1\mathbf{b}_1+h_2\mathbf{b}_2+h_3\mathbf{b}_3)$ go from the origin to the lattice points in reciprocal space so

$$e^{i\mathbf{K}_h\cdot(\mathbf{r}+\mathbf{R}_n)} = e^{i\mathbf{K}_h\cdot\mathbf{r}}. \tag{1.6}$$

The values of k in each direction in the first Brillouin zone are restricted to

$$-\frac{\pi}{a} < k \le \frac{\pi}{a}, \tag{1.7}$$

where a is the unit cell dimension in real space and the end points for k are on the zone surface. The product of the reciprocal space vector in (1.5) with a unit cell vector in real space is

$$\vec{K} \cdot \vec{a}_i = 2\pi h_i, \tag{1.8}$$

where h_i is an integer. The function in (1.2) with the periodicity of the lattice can then be written as

$$u(\vec{r}) = \sum_{\vec{K}} C(\vec{K}) e^{i\vec{K}\cdot\vec{r}}. \qquad (1.9)$$

Since the Bravais lattice is invariant with respect to the symmetry elements of its point group, the corresponding reciprocal lattice must also be invariant with respect

Table 1.3 Equivalent lattices in real and reciprocal space

Lattice in real space	Lattice in reciprocal space
Simple triclinic	Simple triclinic
Simple monoclinic	Simple monoclinic
One-face-centered monoclinic	One-face-centered monoclinic
Simple orthorhombic	Simple orthorhombic
Face-centered orthorhombic	Body-centered orthorhombic
Body-centered orthorhombic	Face-centered orthorhombic
One-face-centered orthorhombic	One-face-centered orthorhombic
Simple tetragonal	Simple tetragonal
Body-centered tetragonal	Body-centered tetragonal
Simple hexagonal	Simple hexagonal
Rhombohedral	Rhombohedral
Simple cubic	Simple cubic
Face-centered cubic	Body-centered cubic
Body-centered cubic	Face-centered cubic

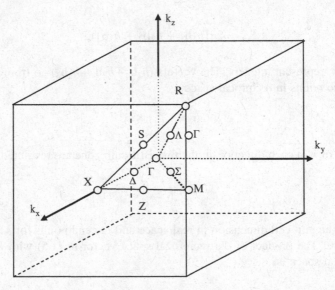

Fig. 1.10 Brillouin zone for a simple cubic lattice with space group O_h^1 in ordinary space. The points with special symmetry are shown

to this point group. Thus the Brillouin zone for this lattice will belong to the same crystal system as the lattice in real space but it can have a different distribution pattern. For example, a face-centered cubic lattice in real space has a body-centered cubic lattice in reciprocal space. Table 1.3 lists the equivalent lattices in real and reciprocal space [7].

An example of a Brillouin zone is shown in Fig. 1.10. This is the reciprocal lattice cell for a simple cubic space group O_h^1 in ordinary space. The points of special symmetry are labeled in Fig. 1.10. The point group symmetry at these points are: O_h for Γ and R; D_{4h} for M and X; C_{4v} for Δ and T; C_{3v} for Λ; and C_{2v} for Z, Σ, and S.

A second example of a Brillouin zone is shown in Fig. 1.11. This is the reciprocal lattice for a hexagonal crystal system. Again the points of special symmetry are shown in Fig 1.11. The Brillouin zone structures for all of the crystal systems are given in [3].

Translational symmetry and reciprocal space are especially important in considering wave-like quasiparticles in crystals. Examples of this for phonons and electrons are given in Chaps. 7 and 8, respectively.

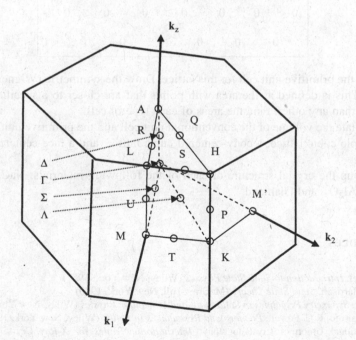

Fig. 1.11 Brillouin zone for a hexagonal crystal structure

1.4   Problems

1. There are 10 point groups in two dimensions with the following symmetry operations:

C_1: E	C_{1v}: E, σ
C_2: E, C_2	C_{2v}: $E, C_2, \sigma, \sigma C_2$
C_3: E, C_3, C_3^2	C_{3v}: $E, C_3, C_3^2, s, sC_3, sC_3^2$
C_4: E, C_4, C_2, C_4^3	C_{4v}: $E, C_4, C_2, C_4^3, s, sC_4, sC_2, sC_4^3$
C_6: $E, C_6, C_3, C_2, C_3^2, C_6^5$	C_{6v}: $E, C_6, C_3, C_2, C_3^2, C_6^5, \sigma, \sigma C_6, \sigma C_3, \sigma C_2, \sigma C_3^2, \sigma C_6^5$

   Draw two-dimensional figures representing each of these symmetry groups.
2. Draw stereograms for each of the ten two-dimensional point groups listed above. Show how an elongating distortion of the C_{4v} shape in problem 1 removes some of the symmetry elements in the stereogram for this point group and leaves a new stereogram for a different point group.
3. A two-dimensional hexagonal lattice is shown in the picture below.

0	0	0	0	0	0	0	
	0	0	0	0	0	0	0
0	0	0	0	0	0	0	
	0	0	0	0	0	0	0

   Draw the primitive unit cell for this lattice. Draw the symmetric (Wigner–Seitz) cell. This is defined as the area with points that are closer to a specific lattice point than any other. Find the areas of each type of cell.
4. Calculate the volume of the conventional unit cell and the primitive unit cell for a simple cubic lattice, a body-centered cubic lattice, and a face-centered cubic lattice.
5. Look up the crystal structure of each of the following materials: NaCl, CsCl, ZnS, Al_2O_3, and Diamond.

References

1. C. Kittel, *Introduction to Solid State Physics* (Wiley, New York, 1957)
2. W.A. Harrison, *Solid State Theory* (McGraw-Hill, New York, 1970)
3. M. Lax, *Symmetry Principles in Solid State and Molecular Physics* (Wiley, New York, 1974)
4. B. DiBartolo, R.C. Powell, *Phonons and Resonances in Solids* (Wiley, New York, 1976)
5. International Union of Crystallography, *International Tables for X-Ray Crystallography* (Kynoch, Birmingham, 1952)
6. M. Sachs, *Solid State Theory* (McGraw-Hill, New York, 1963)
7. J.C. Slater, *Symmetry and Energy Bands in Crystals* (Dover, New York, 1972)

Chapter 2
Group Theory

The formal mathematical treatment of the symmetry of physical systems discussed in Chap.1 is called *group theory*. This chapter summarizes the fundamental properties of group theory that will be used to treat physical examples of symmetry in the succeeding chapters. This book is focused on the practical use of group theory and does not attempt to cover derivations of the fundamental postulates or advanced aspects of this topic. For a rigorous treatment of group theory the reader is referred to [1].

A group is defined as a collection of elements that obey certain criteria and are related to each other through a specific rule of interaction. The rule of interaction is referred to generically as the "multiplication" of two elements. However, the interaction may not be the normal multiplication of two numbers since the elements of a group may not be simple numbers. The number of elements in group h is called the *order of the group*. There are four requirements for a set of elements to form a group:

1. One element, designated E and called the identity element, commutes with all the other elements of the group and multiplication of an element by E leaves the element unchanged. That is, $EA=AE=A$.
2. The result of multiplying any two elements in a group (including the product of an element with itself) is an element of the group. That is, $AB=C$ where A, B, and C are all elements of the group.
3. Every element of the group must have a reciprocal element that is also an element of the group. That is, $AR=RA=E$ where A is an element of the group, R is its reciprocal, and E the identity element and R and E are both members of the group.
4. The associative law of multiplication is valid for the product of any three elements of the group. That is, $A(BC)=(AB)C$.

It is not necessary for the products of elements of a group to obey the commutative law. That is, the element resulting in the product AB may not be the same as the element resulting in the product BA. If the elements of a specific group happen to obey the commutative law the group is said to be *Abelian*.

R.C. Powell, *Symmetry, Group Theory, and the Physical Properties of Crystals*, Lecture Notes in Physics 824, DOI 10.1007/978-1-4419-7598-0_2, © Springer Science+Business Media, LLC 2010

Group multiplication Table

	E	A	B	C	D	F
E	E	A	B	C	D	F
A	A	E	D	F	B	C
B	B	F	E	D	C	A
C	C	D	F	E	A	B
D	D	C	A	B	F	E
F	F	B	C	A	E	D

The properties of a group discussed above can be exemplified in a group multiplication table. Consider a group consisting of six elements represented by the letters A, B, C, D, E, and F that obey the multiplication table shown above. The elements in the table are the product of the element designating its column and the element designating its row. Following this convention, the table shows that the identity element is a member of the group, the product of any two elements is an element of the group, and each group element has an element in the group that is its inverse. Each element appears only once in any given row or column. The associative law holds but the commutative law does not hold for all products so the group is not Abelian. The order of the group is 6.

The multiplication table is useful in identifying subgroups within the whole group. These are subsets of the total set of group elements that meet the requirements of being a group without requiring the other elements of the total group. By inspection, it can be seen that the elements D, E, and F form a subgroup of order 3. Also there are three subgroups of order 2: E,A; E,B; and E,C. Of course the element E by itself always forms a subgroup of order 1. Note that the orders of the subgroups are integral factors of the order of the total group.

Another useful concept in dealing with a group is organizing its elements in conjugate pairs through the use of a similarity transformation. To find the conjugate of an element A, the triple product of A with another element of the group and its reciprocal element is formed. For example,

$$B = X^{-1}AX.$$

This type of product is a similarity transformation, and the elements A and B are said to be conjugates of each other. Every element is conjugate with itself. Also, if A is conjugate with two elements B and C then B and C are conjugate with each other. A complete set of elements that are conjugate to each other form a *class* of the group.

From the multiplication table of the group of elements A, B, C, D, E, F shown above, it is easily seen that E by itself forms a class of order 1. The elements A, B, C form a class of order 3. This can be seen by taking all possible similarity transformations on element A, which gives

$$E^{-1}AE = A, \quad A^{-1}AA = A, \quad B^{-1}AB = C, \quad C^{-1}AC = B, \quad D^{-1}AD = B,$$

$$F^{-1}AF = C,$$

and then doing the same for elements B and C. Similarly, taking all possible similarity transformations on elements D and F show that they form a class of order 2. Note that it is always true that the order of a class is an integral factor of the order of the group.

The type of group of interest here is a *symmetry group*. The elements of this type of group are a complete set of relevant symmetry operations that obey the rules of a group. The specific symmetry groups of interest are those defining the crystal classes discussed in Chap. 1.

2.1 Basic Concepts of Group Theory

The basic concepts of group theory can be demonstrated by considering the spatial symmetry of an object with a specific geometrical shape. The way such an object is transformed by operations about a specific point in space is referred to as *point group* symmetry. The symmetry operations for point groups include rotations about axes, reflections through planes, inversion through a central point, and combinations of these.

By convention, different types of symmetry elements have specific designations [1–4]. To reiterate the designations listed in Chap.1, the identity operation, describing the situation where no transformation takes place, is designated as E. Rotation about an axis of symmetry is designated by C_n which indicates that the object is spatially identical after a rotation of $2\pi/n$ about this axis. For example, a rotation of 180° is represented by the twofold symmetry operation C_2 while a fourfold symmetry axis C_4 represents a rotation of 90°. Since n rotations of C_n take the object back to its original position, $C_n^n = E$. If a reflection plane is perpendicular to the highest order symmetry axis, it is designated by σ_h. If the reflection plane contains the highest order symmetry axis, it is designated by σ_v. Mirror planes diagonal to the rotation axes are designated as σ_d. Mirror operations take twice result in E. For an object possessing a center of symmetry, the inversion operation is designated by i and $i^2 = E$. There are also combined operations. For example the inversion operation is a combined rotation and reflection, $i = C_2\sigma_h$. A combined rotation–reflection operation with the mirror plane perpendicular to the rotation axis is called an improper rotation and designated by S_n. Thus, $S_n = \sigma_h C_n$. The order of successive symmetry operations is important since not all of them commute.

As discussed above, it is convenient to organize the elements of a group into *classes* where all elements in the same class are related to each other by a unitary transformation of another operator of the group. For example, if $T^{-1}AT = A'$ where all of these are elements of the group, A and A' are members of the same class. As stated before, the order of a class must be an integral factor of the order of the group.

The action of the elements of a symmetry group on the physical properties of a system is described in terms of mathematical transformations. The physical properties may be expressed as vectors, matrices, or tensors of higher rank as discussed in

Chap. 3. These form vector spaces, and their transformations in this vector space are the image of the symmetry transformations in coordinate space. For example, the state vectors of a quantum mechanical system transform into each other in the same way as the symmetry transformations of the coordinates describing the system. When the mathematical description of the physical properties of a system transform in the same way as a symmetry group, they are said to be a *representation* of that group. The symmetry elements act as linear operators to produce transformations in a specific representation of the group. A group will have a number of different types of representations associated with different physical properties.

Every group has a one-dimensional *trivial representation* consisting of assigning the number one to all elements of the group. In general, a set of matrices of a specific dimension are assigned to the elements of the group to make a representation of the group. These matrices must obey the same multiplication table as the elements of the group. The matrix of a representation is square and the number of elements in a row or column is the *dimension* of the representation, which is equal to its degeneracy. It is possible to construct many different representations of this type for the same group.

It is always possible to find a similarity transformation that puts a matrix into a box diagonal form

$$A' = T'AT = \begin{pmatrix} [A_1] & & 0 \\ & [A_2] & \\ 0 & & [A_3] \end{pmatrix}. \tag{2.1}$$

In this case A and A' are matrices representing *reducible representations* while the A_i are matrices represent *irreducible representations*. The sum of the squares of the dimensions of the irreducible representations of a group is equal to the order of the group:

$$\sum_i d_i^2 = h. \tag{2.2}$$

The number of irreducible representations of a group is equal to the number of classes in the group.

The spatial position of an object is represented by vectors in Cartesian coordinates. A transformation of the object can be represented by a transformation of these coordinate vectors. The object moves from a vector position designated by the coordinates (x,y,z) to a new position designated by the coordinates (x', y', z') as shown in Fig. 2.1. Any vector $\mathbf{r}$ can be expressed in terms of its Cartesian coordinates using the unit vectors, $\hat{x}, \hat{y}$, and $\hat{z}$. A transformation operation can then be applied to each component and the new components recombined to give the transformed vector $\mathbf{r}'$. If a rotation about the major symmetry axis (usually taken to be the z-axis) is designated by an angle θ, the transformation is given as

Fig. 2.1 Transformation of
Cartesian coordinates

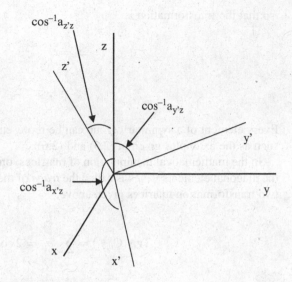

$$x' = x \cos \theta + y \sin \theta$$
$$y' = -x \sin \theta + y \cos \theta . \qquad (2.3)$$
$$z' = z$$

In matrix form this coordinate transformation is written as

$$\begin{pmatrix} x' \\ y' \\ z' \end{pmatrix} = \vec{\vec{A}} \begin{pmatrix} x \\ y \\ z \end{pmatrix} = \begin{pmatrix} a_{x'x} & a_{x'y} & a_{x'z} \\ a_{y'x} & a_{y'y} & a_{y'z} \\ a_{z'x} & a_{z'y} & a_{z'z} \end{pmatrix} \begin{pmatrix} x \\ y \\ z \end{pmatrix} . \qquad (2.4)$$

The matrix elements $a_{i'j}$ are the direction cosines of the coordinate represented by i' with respect to the coordinate represented by j as shown in Fig. 2.1. For the example of a rotation about the z-axis given by (2.3) the transformation matrix is

$$\vec{\vec{A}}(C_{\theta(z)}) = \begin{pmatrix} \cos \theta & \sin \theta & 0 \\ -\sin \theta & \cos \theta & 0 \\ 0 & 0 & 1 \end{pmatrix} . \qquad (2.5)$$

A symmetry operation consisting of a mirror reflection plane perpendicular to the z-axis would be represented by the matrix

$$\vec{\vec{A}}(\sigma_h) = \begin{pmatrix} 1 & 0 & 0 \\ 0 & 1 & 0 \\ 0 & 0 & -1 \end{pmatrix} , \qquad (2.6)$$

so that the transformation is

$$x' = x$$
$$y' = y \qquad (2.7)$$
$$z' = -z.$$

Every element of a symmetry group can be represented by a transformation matrix such as the examples given in (2.5) and (2.6).

In the mathematical manipulation of matrices, one useful property is the sum of the diagonal elements which is called the *trace* of the matrix. In the examples of the two transformation matrices given above,

$$\mathbf{Tr}\vec{\mathbf{A}}(C_{\theta(z)}) = \sum_i a_{ii} = 2\cos\theta + 1 \qquad (2.8)$$

and

$$\mathbf{Tr}\vec{\mathbf{A}}(\sigma_h) = \sum_{ii} a_{ii} = 1. \qquad (2.9)$$

The trace of a transformation matrix representing a symmetry operation is called the *character* of the operation in that representation and is designated by χ.

Characters of matrix operators have special properties that make them useful working with group theory.

1. Since the trace of a matrix is invariant under a similarity transformation, all symmetry operations belonging to the same class of the group have the same character.
2. The character of a reducible representation is equal to the sum of the characters of the irreducible representations that it contains.
3. The number of times that a specific irreducible representation is contained in the reduction of a reducible representation can be determined by

$$n^{(i)} = \frac{1}{h} \sum_A \chi_A^{(i)} \chi_A. \qquad (2.10)$$

Here $\chi_A^{(i)}$ is the character of the operation A in the ith irreducible representation while χ_A is the character of the same operation in the reducible representation. The sum is over all of the symmetry operations of the group of order h.

4. For any irreducible representation, the sum of the squares of the characters of all the operations equals the order of the group

$$\sum_A \chi_i^2(A) = h. \tag{2.11}$$

5. The set of characters for two different irreducible representations are orthogonal

$$\sum_A \chi_i(A)\chi_j(A) = 0, \quad i \neq j. \tag{2.12}$$

6. The *direct product* of two representations is found by multiplying the characters of a specific operation in these two representations to give the character of that operation in the product representation. A direct product representation is usually a reducible representation of the group.

Another important property of transformation matrices is that irreducible representations are orthogonal and obey the relationship

$$\sum_A [\Gamma_i(A)_{mn}][\Gamma_j(A)_{m'n'}]^* = \frac{h}{\sqrt{d_i d_j}} \delta_{ij}\delta_{mm'}\delta_{nn'}. \tag{2.13}$$

Here $\Gamma_i(A)_{mn}$ is the mn matrix element of the transformation matrix for operation A in the Γ_i irreducible representation.

Each representation of a symmetry group operates on a set of functions that transform into each other under that representation of the group. These are *called basis functions* for that representation. For physical systems they represent a specific physical property of the system. In the example of the coordinate transformation discussed above, the vector coordinates x, y, and z are the set of basis functions. Any property described by a vector will transform like this set of basis functions according to the representation of the group of symmetry elements for the system. The rotation axes R_x, R_y, and R_z can also act as a set of basis functions for irreducible representations of a group. These differ from the spatial coordinates because a symmetry operation may change the direction of rotation. A third common set of basis functions are the six components of a pseudovector arising from a vector product. These basis functions are discussed in the examples given below, and in Chap. 4 it is shown how spherical harmonic functions can also be used as basis functions.

2.2 Character Tables

A *character table* for a symmetry group lists the characters for each class of operations in the group for each of the irreducible representations of the group. The character tables for each of the 32 crystallographic point groups discussed in Chap. 1 are given in Tables 2.1–2.32. These are very useful in the application of group theory to determine the properties of crystals [4–6].

Table 2.1 Character table for point group C_1

C_1	E
A	1

Table 2.2 Character table for point group C_s

C_s	E	σ_h	Basis components		
A'	1	1	$x,y,$	R_z	x^2,y^2z^2,xy
A''	1	-1	$z,$	R_x,R_y	yz,xz

Table 2.3 Character table for point group C_i

C_i	E	i	Basis components		
A_g	1	1		R_x,R_y,R_z	x^2,y^2,z^2,xy,xz,yz
A_u	1	-1	x,y,z		

Table 2.4 Character table for point group for C_2

C_2	E	C_2	Basis components		
A	1	1	z	R_z	x^2,y^2,z^2, xy
B	1	-1	x,y	R_x,R_y	yz,xz

Table 2.5 Character table for point group C_{2h}

C_{2h}	E	C_2	i	σ_h	Basis components		
A_g	1	1	1	1		R_z	x^2,y^2,z^2, xy
B_g	1	-1	1	-1		R_x,R_y	yz,xz
A_u	1	1	-1	-1	z		
B_u	1	-1	-1	1	x,y		

Table 2.6 Character table for point group C_{2v}

C_{2v}	E	C_2	$\sigma_v(xz)$	$\sigma_v'(yz)$	Basis components		
A_1	1	1	1	1	z		x^2,y^2,z^2
A_2	1	1	-1	-1		R_z	xy
B_1	1	-1	1	-1	x	R_y	xy
B_2	1	-1	-1	1	y	R_x	yz

Table 2.7 Character table for point group D_2

D_2	E	$C_2(z)$	$C_2(y)$	$C_2(x)$	Basis components		
A	1	1	1	1			x^2,y^2,z^2
B_1	1	1	-1	-1	z	R_z	xy
B_2	1	-1	1	-1	y	R_y	xz
B_3	1	-1	-1	1	x	R_x	yz

Table 2.8 Character table for point group D_{2h}

D_{2h}	E	$C_2(z)$	$C_2(y)$	$C_2(x)$	i	$\sigma(xy)$	$\sigma(xz)$	$\sigma(yz)$	Basis components		
A_g	1	1	1	1	1	1	1	1			x^2,y^2,z^2
B_{1g}	1	1	-1	-1	1	1	-1	-1	R_z		xy
B_{2g}	1	-1	1	-1	1	-1	1	-1	R_y		xz
B_{3g}	1	-1	-1	1	1	-1	-1	1	R_x		yz
A_u	1	1	1	1	-1	-1	-1	-1			
B_{1u}	1	1	-1	-1	-1	-1	1	1	z		
B_{2u}	1	-1	1	-1	-1	1	-1	1	y		
B_{3u}	1	-1	-1	1	-1	1	1	-1	x		

Table 2.9 Character table for point group C_4

C_4	E	C_4	C_2	C_4^3	Basis components		
A	1	1	1	1	z	R_z	x^2+y^2,z^2
B	1	-1	1	-1			x^2-y^2,xy
E^*	2	0	-2	0	(x,y)	(R_x,R_y)	(yz,xz)

Table 2.10 Character table for point group C_{4h}

C_{4h}	E	C_4	C_2	C_4^3	i	S_4^3	σ_h	S_4	Basis components		
A_g	1	1	1	1	1	1	1	1		R_z	x^2+y^2,z^2
B_g	1	-1	1	-1	1	-1	1	-1			x^2-y^2,xy
E_g^*	2	0	-2	0	2	0	-2	0		(R_x,R_y)	(yz,xz)
A_u	1	1	1	1	-1	-1	-1	-1	z		
B_u	1	-1	1	-1	-1	1	-1	1			
E_u^*	2	0	-2	0	2	0	-2	0	(x,y)		

Table 2.11 Character table for point group C_{4v}

C_{4v}	E	$2C_4$	C_2	$2\sigma_v$	$2\sigma_d$	Basis components		
A_1	1	1	1	1	1	z		x^2+y^2,z^2
A_2	1	1	1	-1	-1		R_z	
B_1	1	-1	1	1	-1			x^2-y^2
B_2	1	-1	1	-1	1			xy
E	2	0	-2	0	0	(x,y)	(R_x,R_y)	(yz,xz)

Table 2.12 Character table for point group S_4

S_4	E	S_4	C_2	S_4^3	Basis components		
A	1	1	1	1		R_z	x^2+y^2,z^2
B	1	-1	1	-1	z		x^2-y^2,xy
E^*	2	0	-2	0	(x,y)	(R_x,R_y)	(yz,xz)

Table 2.13 Character table for point group D_4

D_4	E	$2C_4$	C_2	$2C_2'$	$2C_2''$	Basis components		
A_1	1	1	1	1	1			x^2+y^2,z^2
A_2	1	1	1	-1	-1	z	R_z	
B_1	1	-1	1	1	-1			x^2-y^2
B_2	1	-1	1	-1	1			xy
E	2	0	-2	0	0	(x,y)	(R_x,R_y)	(yz,xz)

Table 2.14 Character table for point group D_{4h}

D_{4h}	E	$2C_4$	C_2	$2C_2'$	$2C_2''$	i	$2S_4$	σ_h	$2\sigma_v$	$2\sigma_d$	Basis components		
A_{1g}	1	1	1	1	1	1	1	1	1	1			x^2+y^2,z^2
A_{2g}	1	1	1	-1	-1	1	1	1	-1	-1		R_z	
B_{1g}	1	-1	1	1	-1	1	-1	1	1	-1			x^2-y^2
B_{2g}	1	-1	1	-1	1	1	-1	1	-1	1			xy
E_g	2	0	-2	0	0	2	0	-2	0	0		(R_x,R_y)	(yz,xz)
A_{1u}	1	1	1	1	1	-1	-1	-1	-1	-1			
A_{2u}	1	1	1	-1	-1	-1	-1	-1	1	1	z		
B_{1u}	1	-1	1	1	-1	-1	1	-1	-1	1			
B_{2u}	1	1	1	-1	1	-1	1	-1	1	-1			
E_u	2	0	-2	0	0	-2	0	2	0	0	(x,y)		

Table 2.15 Character table for point group D_{2d}

D_{2d}	E	$2S_4$	C_2	$2C_2'$	$2\sigma_d$	R	$2RS_4$	RC_2	$2RC_2'$	$2R\sigma_d$	Basis components		
A_1	1	1	1	1	1	1	1	1	1	1			x^2+y^2,z^2
A_2	1	1	1	-1	-1	1	1	1	-1	-1		R_z	
B_1	1	-1	1	1	-1	1	-1	1	1	-1			x^2-y^2
B_2	1	-1	1	-1	1	1	-1	1	-1	1	z		xy
E	2	0	-2	0	0	2	0	-2	0	0	(x,y) (R_x,R_y)		(yz,xz)
$D_{1/2}$	2	$\sqrt{2}$	0	0	0	-2	$-\sqrt{2}$	0	0	0			
$_2S$	2	$-\sqrt{2}$	0	0	0	-2	$\sqrt{2}$	0	0	0			

Table 2.16 Character table for point group C_3

C_3	E	C_3	C_3^2	Basis components		
A	1	1	1	z	R_z	x^2+y^2,z^2
$E*$	2	-1	-1	(x,y)	(R_x,R_y)	$(x_2-y_2,xy)(yz,xz)$

Table 2.17 Character table for point group C_{3v}

C_{3v}	E	$2C_3$	$3\sigma_v$	Basis components		
A_1	1	1	1	z		x^2+y^2,z^2
A_2	1	1	-1		R_z	
E	2	-1	0	(x,y)	(R_x,R_y)	$(x^2-y^2,xy)(yz,xz)$

Table 2.18 Character table for point group C_{3h}

C_{3h}	E	C_3	C_3^2	σ_h	S_3	S_3^5	Basis components		
A'	1	1	1	1	1	1		R_z	x^2+y^2,z^2
$E'*$	2	-1	-1	2	-1	-1	(x,y)		(x^2-y^2,xy)
A''	1	1	1	-1	-1	-1	z		
$E''*$	2	-1	-1	-2	1	1		(R_x,R_y)	(yz,xz)

Table 2.19 Character table for point group D_3

D_3	E	$2C_3$	$3C_2$	Basis components		
A_1	1	1	1			x^2+y^2,z^2
A_2	1	1	-1	z	R_z	
E	2	-1	0	(x,y)	(R_x,R_y)	$(x^2-y^2,xy)(yz,xz)$

Table 2.20 Character table for point group D_{3d}

D_{3d}	E	$2C_3$	$3C_2$	i	$2S_6$	$3\sigma_d$	Basis components		
A_{1g}	1	1	1	1	1	1			x^2+y^2,z^2
A_{2g}	1	1	-1	1	1	-1		R_z	
E_g	2	-1	0	2	-1	0		(R_x,R_y)	$(x^2-y^2,xy)(yz,xz)$
A_{1u}	1	1	1	-1	-1	-1			
A_{2u}	1	1	-1	-1	-1	1	z		
E_u	2	-1	0	-2	1	0	(x,y)		

Table 2.21 Character table for point group S_6

S_6	E	C_3	C_3^2	i	S_6^5	S_6	Basis components		
A_g	1	1	1	1	1	1		R_z	x^2+y^2,z^2
E_g^*	2	-1	-1	2	-1	-1		(R_x,R_y)	$(x^2-y^2,xy)(yz,xz)$
A_u	1	1	1	-1	-1	-1	z		
E_u^*	2	-1	-1	-2	1	1	(x,y)		

Table 2.22 Character table for point group C_6

C_6	E	C_6	C_3	C_2	C_3^2	C_6^5	Basis components		
A	1	1	1	1	1	1	z	R_z	x^2+y^2,z^2
B	1	-1	1	-1	1	-1			
E_1^*	2	1	1	-2	1	1	(x,y)	(R_x,R_y)	(xz,yz)
E_2^*	2	-1	1	2	1	-1			(x^2-y^2,xy)

Table 2.23 Character table for point group C_{6h}

C_{6h}	E	C_6	C_3	C_2	C_3^2	C_6^5	i	S_3^5	S_6^5	σ_h	S_6	S_3	Basis components	
A_g	1	1	1	1	1	1	1	1	1	1	1	1	R_z	x^2+y^2,z^2
B_g	1	-1	1	-1	1	-1	1	-1	1	-1	1	-1		
E_{1g}^*	2	1	-1	-2	-1	1	2	1	-1	-2	-1	1	(R_x,R_y)	(xz,yz)
E_{2g}^*	2	-1	-1	2	-1	-1	2	-1	-1	2	-1	-1		(x^2-y^2,xy)
A_u	1	1	1	1	1	1	-1	-1	-1	-1	-1	-1	z	
B_u	1	-1	1	-1	1	-1	-1	1	-1	1	-1	1		
E_{1u}^*	2	1	-1	-2	-1	1	-2	-1	1	2	1	-1	(x,y)	
E_{2u}^*	2	-1	-1	2	-1	-1	-2	1	1	-2	1	1		

Table 2.24 Character table for point group C_{6v}

C_{6v}	E	$2C_6$	$2C_3$	C_2	$3\sigma_v$	$3\sigma_d$	Basis components		
A_1	1	1	1	1	1	1	z		x^2+y^2,z^2
A_2	1	1	1	1	-1	-1		R_z	
B_1	1	-1	1	-1	1	-1			
B_2	1	-1	1	-1	-1	1			
E_1	2	1	-1	-2	0	0	(x,y)	(R_x,R_y)	(xz,yz)
E_2	2	-1	-1	2	0	0			(x^2-y^2,xy)

Table 2.25 Character table for point group D_6

D_6	E	$2C_6$	$2C_3$	C_2	$3C_2'$	$3C_2''$	Basis components		
A_1	1	1	1	1	1	1			x^2+y^2,z^2
A_2	1	1	1	1	-1	-1	z	R_z	
B_1	1	-1	1	-1	1	-1			
B_2	1	-1	1	-1	-1	1			
E_1	2	1	-1	-2	0	0	(x,y)	(R_x,R_y)	(xz,yz)
E_2	2	-1	-1	2	0	0			(x^2-y^2,xy)

Table 2.26 Character table for point group D_{6h}

D_{6h}	E	$2C_6$	$2C_3$	C_2	$3C_2'$	$3C_2''$	i	$2S_3$	$2S_6$	σ_h	$3\sigma_d$	$3\sigma_v$	Basis components	
A_{1g}	1	1	1	1	1	1	1	1	1	1	1	1		x^2+y^2,z^2
A_{2g}	1	1	1	1	-1	-1	1	1	1	1	-1	-1	R_z	
B_{1g}	1	-1	1	-1	1	-1	1	-1	1	-1	1	-1		
B_{2g}	1	-1	1	-1	-1	1	1	-1	1	-1	-1	1		
E_{1g}	2	1	-1	-2	0	0	2	1	-1	-2	0	0	(R_x,R_y)	(xz,yz)
E_{2g}	2	-1	-1	2	0	0	2	-1	-1	2	0	0		(x^2-y^2,xy)
A_{1u}	1	1	1	1	1	1	-1	-1	-1	-1	-1	-1		
A_{2u}	1	1	1	1	-1	-1	-1	-1	-1	-1	1	1	z	
B_{1u}	1	-1	1	-1	1	-1	-1	1	-1	1	-1	1		
B_{2u}	1	-1	1	-1	-1	1	-1	1	-1	1	1	-1		
E_{1u}	2	1	-1	-2	0	0	-2	-1	1	2	0	0	(x,y)	
E_{2u}	2	-1	-1	2	0	0	-2	1	1	-2	0	0		

Table 2.27 Character table for point group D_{3h}

D_{3h}	E	$2C_3$	$3C_2$	σ_h	$2S_3$	$3\sigma_v$	Basis components		
A_1'	1	1	1	1	1	1			x^2+y^2,z^2
A_2'	1	1	-1	1	1	-1		R_z	
E'	2	-1	0	2	-1	0	(x,y)		(x^2-y^2,xy)
A_1''	1	1	1	-1	-1	-1			
A_2''	1	1	-1	-1	-1	1	z		
E''	2	-1	0	-2	1	0		(R_x,R_y)	(xz,yz)

Table 2.28 Character table for point group T

T	E	$3C_2$	$4C_3$	$4C_3^2$	Basis Components		
A	1	1	1	1			z^2
E^*	2	2	-1	-1			x^2+y^2, x^2-y^2
T	3	-1	0	0	(x,y,z)	(R_x,R_y,R_z)	(xz,yz, xy)

Table 2.29 Character table for point group T_h

T_h	E	$3C_2$	$4C_3$	$4C_3^2$	i	$3\sigma_h$	$4iC_3$	$4iC_3^2$	Basis Components	
A_g	1	1	1	1	1	1	1	1		$x^2+y^2+z^2$
E_g^*	2	2	-1	-1	2	2	-1	-1		$(2z^2-x^2-y^2,x^2-y^2)$
T_g	3	-1	0	0	3	-1	0	0	(R_x,R_y,R_z)	xz,yz,xy
A_u	1	1	1	1	-1	-1	-1	-1		
E_u^*	2	2	-1	-1	-2	-2	1	1		
T_u	3	-1	0	0	-3	1	0	0	(x,y,z)	

Table 2.30 Character table for point group T_d

T_d	E	$8C_3$	$3C_2$	$6S_4$	$6\sigma_d$	Basis components	
A_1	1	1	1	1	1		$x^2+y^2+z^2$
A_2	1	1	1	-1	-1		
E	2	-1	2	0	0		$(2z^2-x^2-y^2,x^2-y^2)$
T_1	3	0	-1	1	-1	(R_x,R_y,R_z)	
T_2	3	0	-1	-1	1	(x,y,z)	(xz,yz,xy)

Table 2.31 Character table for point group O

O	E	$8C_3$	$6C_2$	$6C_4$	$3C_4^2$	Basis components	
A_1	1	1	1	1	1		$x^2+y^2+z^2$
A_2	1	1	-1	-1	1		
E	2	-1	0	0	2		$(2z^2-x^2-y^2,x^2-y^2)$
T_1	3	0	-1	1	-1	(x,y,z)	(R_x,R_y,R_z)
T_2	3	0	1	-1	-1		(xz,yz,xy)

Table 2.32 Character table for point group O_h

O_h	E	$8C_3$	$6C_2$	$6C_4$	$3C_4^2$	i	$6S_4$	$8S_6$	$3\sigma_h$	$6\sigma_d$	Basis components
A_{1g}	1	1	1	1	1	1	1	1	1	1	$x^2+y^2+z^2$
A_{2g}	1	1	-1	-1	1	1	-1	1	1	-1	
E_g	2	-1	0	0	2	2	0	-1	2	0	$(2z^2-x^2-y^2,$ $x^2-y^2)$
T_{1g}	3	0	-1	1	-1	3	1	0	-1	-1	(R_x,R_y,R_z)
T_{2g}	3	0	1	-1	-1	3	-1	0	-1	1	(xz,yz,xy)
$D_{1/2g}$	2	1	0	$\sqrt{2}$	0	2	$\sqrt{2}$	1	0	0	
$_2S_g$	2	1	0	$-\sqrt{2}$	0	2	$-\sqrt{2}$	1	0	0	
$D_{3/2g}$	4	-1	0	0	0	4	0	-1	0	0	
A_{1u}	1	1	1	1	1	-1	-1	-1	-1	-1	
A_{2u}	1	1	-1	-1	1	-1	1	-1	-1	1	
E_u	2	-1	0	0	2	-2	0	1	-2	0	
T_{1u}	3	0	-1	1	-1	-3	-1	0	1	1	(x,y,z)
T_{2u}	3	0	1	-1	-1	-3	1	0	1	-1	
$D_{1/2u}$	2	1	0	$\sqrt{2}$	0	-2	$-\sqrt{2}$	-1	0	0	
$_2S_u$	2	1	0	$-\sqrt{2}$	0	-2	$\sqrt{2}$	-1	0	0	
$D_{3/2u}$	4	-1	0	0	0	-4	0	1	0	0	

Table 2.32 Character table for point group O_h (continued)

O_h	R	$8RC_3$	$6RC_2$	$6RC_4$	$3RC_4^2$	Ri	$6RS_4$	$8RS_6$	$3R\sigma_h$	$6R\sigma_d$
A_{1g}	1	1	1	1	1	1	1	1	1	1
A_{2g}	1	1	-1	-1	1	1	-1	1	1	-1
E_g	2	-1	0	0	2	2	0	-1	2	0
T_{1g}	3	0	-1	1	-1	3	1	0	-1	-1
T_{2g}	3	0	1	-1	-1	3	-1	0	-1	1
$D_{1/2g}$	-2	-1	0	$-\sqrt{2}$	0	-2	$-\sqrt{2}$	-1	0	0
$_2S_g$	-2	-1	0	$\sqrt{2}$	0	-2	$\sqrt{2}$	-1	0	0
$D_{3/2g}$	-4	1	0	0	0	-4	0	1	0	0
A_{1u}	1	1	1	1	1	-1	-1	-1	-1	-1
A_{2u}	1	1	-1	-1	1	-1	1	-1	-1	1
E_u	2	-1	0	0	2	-2	0	1	-2	0
T_{1u}	3	0	-1	1	-1	-3	-1	0	1	1
T_{2u}	3	0	1	-1	-1	-3	1	0	1	-1
$D_{1/2u}$	-2	-1	0	$-\sqrt{2}$	0	2	$\sqrt{2}$	1	0	0
$_2S_u$	-2	-1	0	$\sqrt{2}$	0	2	$-\sqrt{2}$	1	0	0
$D_{3/2u}$	-4	1	0	0	0	4	0	-1	0	0

There are different notations used to designate irreducible representations in group theory. Γ is used for a generic representation. The character tables shown here use the Mulliken notation which distinguishes between different types of irreducible representations. One-dimensional representations are designated by either A or B. The former is used when the character of the major rotation operation is 1 and the latter is used if the character of this operation is -1. Two-dimensional irreducible representations are designated by E and three-dimensional representations are designated by T. Subscripts 1 and 2 are used if the representation has symmetric ($\chi(C_2)=1$) or antisymmetric ($\chi(C_2)=-1$) twofold rotations perpendicular to the principal rotation axis or vertical symmetry plane. Primes and double primes are used to indicate symmetric or antisymmetric operations with respect to a horizontal plane of symmetry σ_h. If the group has a center of inversion symmetry, the subscripts g (*gerade*) and u (*ungerade*) are used to designate representations that are symmetric and antisymmetric with respect to this operation, respectively.

For each character table, the point group is designated by its Schoenflies notation in the top left-hand corner. The next part of the top row lists the symmetry elements of the group collected into classes. The final part of this row lists some of the possible basis functions for the irreducible representations. The first column of the character table below the first row lists the irreducible representations of the group in the order of increasing dimensions. The main body of the table lists the characters of the sy mmetry elements in each irreducible representation. The last column shows the components of a vector, rotation, or vector product basis function that transforms according to that specific irreducible representation and therefore acts as a basis for that representation.

In several of the character tables, the two-dimensional E representation is shown with an asterisk, E^*. This is because the characters for this representation are imaginary or complex. Technically they should be decomposed into two different representations whose characters are complex conjugates of each other. Doing this allows for the rule of group theory to be fulfilled that the number of irreducible representations of a group is equal to the number of classes of elements in the group. However in applying group theory to physical problems, the characters need to be real so the sum of the characters of these two complex representations is used for the characters of the real representation. In each case the complex character for a rotation axis of order n is

$$\varepsilon_n^m = \exp(2\pi i/n) = \cos\frac{2\pi m}{n} + i\sin\frac{2\pi m}{n}. \qquad (2.14)$$

Using this expression, $\varepsilon_n^0 = \varepsilon_n^n = 1$, $\varepsilon_n^{n/2} = -1$, and $\varepsilon_n^{n/4} = i$. Thus the double-valued representation E in the point group C_3 is actually two complex representations with sets of characters for the classes E, C_3, and C_3^2 of 1, $(-1/2 + i\sqrt{3}/2)$, $(-1/2 - i\sqrt{3}/2)$ and 1, $(-1/2 - i\sqrt{3}/2)$, $(-1/2 + i\sqrt{3}/2)$. Adding these gives the set of characters for the classes of the E^* irreducible representation $2, -1, -1$. Only the characters of the real representations are listed in the character tables.

If a system is characterized by a function that has half-integer values instead of integer values, it is necessary to work with *double groups* [3, 7–9]. In this case the order of the group increases and the number of irreducible representations increases accordingly. This situation occurs most commonly in dealing with spin or half-integer angular momentum in atomic physics. The spin of an electron is represented by a function that has two orientations with respect to an axis of quantization. The *Pauli spin operators* describing this situation are 2×2 matrices

$$\sigma_x = \begin{bmatrix} 0 & 1 \\ 1 & 0 \end{bmatrix}, \quad \sigma_y = \begin{bmatrix} 0 & -i \\ i & 0 \end{bmatrix}, \quad \sigma_z = \begin{bmatrix} 1 & 0 \\ 0 & -1 \end{bmatrix}. \qquad (2.15)$$

These are related to the angular momentum operator $\mathbf{J}$ by

$$\sigma = 2\mathbf{J}.$$

Following the treatment of $\mathbf{J}$ in quantum mechanics, the angular momentum raising and lowering operators for spin can be expressed in terms of the Pauli spin operators [3].

The Pauli spin operators obey a multiplication table that has the properties of a group. A rotation about an axis n in the two dimensional spin representation is given by the operator

$$R(\varphi, \vec{n}) = e^{-i(1/2)\varphi\sigma\cdot\mathbf{n}} = \cos\frac{1}{2}\varphi - i\sigma\cdot\mathbf{n}\sin\frac{1}{2}\varphi. \qquad (2.16)$$

The operator $R(\varphi, \mathbf{n})$ is also a 2×2 matrix.

An important result of (2.16) is that a rotation of 2π is not the identity operator for the group:

$$R(\varphi + 2\pi, \mathbf{n}) = \cos(\pi + \varphi/2) - i\sigma \cdot \mathbf{n} \sin(\pi + \varphi/2)$$
$$= -\cos(\varphi/2) + i\sigma \cdot \mathbf{n} \sin(\varphi/2) = -R(\varphi, \mathbf{n}).$$

Instead an operator representing a rotation of 4π must be introduced as the identity E while a rotation of 2π is a new operator R. Then R multiplied by all of the other operators of the group gives the additional group operators. This leads to additional irreducible representations.

In group theory, spin is represented by a two-dimensional irreducible representation $\Gamma_{1/2}$. For some spatial operations the characters for C_n and RC_n are different and these are referred to as *double valued*. The complete spatial and spin state of a system is represented by the product of $\Gamma_{1/2}$ with the irreducible representations describing the spatial state of the system. In some cases this direct product results in other new irreducible representations of the group. The character tables for the D_{2d} and O_h groups show examples of the extra elements and irreducible representations associated with double groups. These double-valued representations are discussed in greater detail in Chap. 4 and examples given of how to determine the characters of the half-integer representations. These concepts are especially important for treating magnetic properties and the effects of time reversal in quantum mechanical systems.

The irreducible representations for space groups are discussed in Chap. 8.

2.3 Group Theory Examples

2.3.1 C_{3v} Point Group

The best way to demonstrate the use of group theory is to work out some specific examples. Consider an object with the shape of an equilateral triangle as shown in Fig. 2.2. By inspection, this object has six symmetry elements: the identity E; rotation by 120° around the z-axis C_3; rotation by 240° around the z-axis C_3^2; mirror reflection through the plane containing the y and z axes σ_1; and mirror reflections through the planes containing the z-axis and either axis 2 or axis 3, designated σ_2 and σ_3, respectively. Therefore the order of the group is 6. These elements can be displayed in a multiplication table as shown in Table 2.33. This shows that the product of any two elements is an element of the group. It also shows that every element of the group has a reciprocal element that is an element of the group. It also shows that the associative law of multiplication holds for these elements. Thus all of the criteria for being a group have been met.

The multiplication rules shown in Table 2.33 can be used to apply similarity transformations to these elements which allow them to be grouped into classes:

$EC_3E=C_3,$	$EC_3^2E = C_3^2,$
$C_3^2C_3C_3 = C_3,$	$C_3^2C_3^2C_3 = C_3^2,$
$C_3C_3C_3^2 = C_3,$	$C_3C_3^2C_3^2 = C_3^2,$
$\sigma_1C_3\sigma_1 = C_3^2,$	$\sigma_1C_3^2\sigma_1 = C_3,$
$\sigma_2C_3\sigma_2 = C_3^2,$	$\sigma_2C_3^2\sigma_2 = C_3,$
$\sigma_3C_3\sigma_3 = C_3^2,$	$\sigma_3C_3^2\sigma_3 = C_3.$

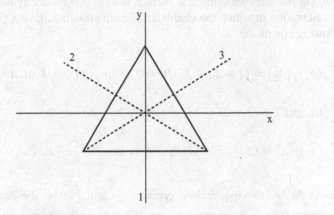

Fig. 2.2 Equilateral triangle. The z-axis direction is out of the page

Table 2.33 Multiplication table for equilateral triangle symmetry elements

	E	C_3	C_3^2	σ_1	σ_2	σ_3
E	E	C_3	C_3^2	σ_1	σ_2	σ_3
C_3	C_3	C_3^2	E	σ_3	σ_1	σ_2
C_3^2	C_3^2	E	C_3	σ_2	σ_3	σ_1
σ_1	σ_1	σ_2	σ_3	E	C_3	C_3^2
σ_2	σ_2	σ_3	σ_1	C_3^2	E	C_3
σ_3	σ_3	σ_1	σ_2	C_3	C_3^2	E

This shows that the elements C_3 and C_3^2 form one class having two symmetry elements. Proceeding in the same way for the three mirror planes show that the elements $\sigma_1, \sigma_2,$ and σ_3 form another class. Also, the element E forms a class by itself.

Since there are three classes in this group there must be three irreducible representations for the group and the sum of the squares of their dimensions must equal the order of the group, 6. This is only possible if there are two one-dimensional irreducible representations and one two-dimensional irreducible representation. There are two ways to develop the character table for these representations. The first is to express the character table in terms of unknown characters and then use the orthogonality of irreducible representations to calculate the characters. In this case the three classes and three irreducible representations can be written as

	E	$2C_3$	3σ
A_1	1	1	1
A_2	1	a	b
E	2	c	d

which reflect the fact that the character of the identity operation is always the dimension of the representation and there is always one totally symmetric irreducible representation in which the character of each class is 1. Using (2.11) provides the following equations:

$$\frac{1}{6}\sum_r \gamma(A_1^2)\gamma(A_2^2) = (1 + 2a + 3b)/6 = 0 \therefore 2a + 3b = -1 \text{ so } a = 1, b = -1,$$

$$\frac{1}{6}\sum_r \gamma(A_1)\gamma(E) = (2 + 2c + 3d)/6 = 0 \therefore 2c + 3d = -2,$$

$$\frac{1}{6}\sum_r \gamma(A_2)\gamma(E) = (2 + 2c - 3d)/6 = 0 \therefore 2c - 3d = -2.$$

Combining the last two expressions gives $c=-1$ and $d=0$, so the character table is

	E	$2C_3$	3σ
A_1	1	1	1
A_2	1	1	-1
E	2	-1	0

The second way to derive the character table for this group is to consider how the Cartesian coordinates transform under the elements of the group. In this case

$$\begin{pmatrix} x' \\ y' \\ z' \end{pmatrix} = E \begin{pmatrix} x \\ y \\ z \end{pmatrix} = \begin{pmatrix} x \\ y \\ z \end{pmatrix} \quad \therefore E = \begin{pmatrix} 1 & 0 & 0 \\ 0 & 1 & 0 \\ 0 & 0 & 1 \end{pmatrix},$$

$$\begin{pmatrix} x' \\ y' \\ z' \end{pmatrix} = C_3 \begin{pmatrix} x \\ y \\ z \end{pmatrix} = \begin{pmatrix} -\frac{1}{2}x + \frac{\sqrt{3}}{2}y \\ -\frac{\sqrt{3}}{2}x - \frac{1}{2}y \\ z \end{pmatrix} \quad \therefore C_3 = \begin{pmatrix} -\frac{1}{2} & \frac{\sqrt{3}}{2} & 0 \\ -\frac{\sqrt{3}}{2} & -\frac{1}{2} & 0 \\ 0 & 0 & 1 \end{pmatrix},$$

$$\begin{pmatrix} x' \\ y' \\ z' \end{pmatrix} = \sigma_1 \begin{pmatrix} x \\ y \\ z \end{pmatrix} = \begin{pmatrix} -x \\ y \\ z \end{pmatrix} \quad \therefore \sigma_1 = \begin{pmatrix} -1 & 0 & 0 \\ 0 & 1 & 0 \\ 0 & 0 & 1 \end{pmatrix}.$$

This is a reducible representation Γ that has characters given in the following table: The final line in this table shows the reduction of the representation Γ in terms of the irreducible representations using the expression from (2.10)

	E	$2C_3$	3σ
A_1	1	1	1
A_2	1	1	-1
E	2	-1	0
Γ	3	0	$1 = A_1 + E$

$$n^{(i)} = \frac{1}{h} \sum_A \chi_A^{(i)} \chi_A.$$

For A_1 this gives $n^{(A1)} = (1/6)(3 + 0 + 3) = 1$. For A_2 it gives $n^{(A2)} = (1/6)$ $(3+0-3)=0$. For E it gives $n^{(E)} = (1/6)(6+0+0) = 1$.

The three transformation matrices found above have a box diagonal form

$$E = \begin{pmatrix} 1 & 0 & 0 \\ 0 & 1 & 0 \\ \hline 0 & 0 & 1 \end{pmatrix}, \quad C_3 = \begin{pmatrix} -\frac{1}{2} & \frac{\sqrt{3}}{2} & 0 \\ -\frac{\sqrt{3}}{2} & -\frac{1}{2} & 0 \\ \hline 0 & 0 & 1 \end{pmatrix}, \quad \sigma_1 = \begin{pmatrix} -1 & 0 & 0 \\ 0 & 1 & 0 \\ \hline 0 & 0 & 1 \end{pmatrix}.$$

The boxes in the upper left-hand corner are the matrices for the irreducible representation E while the boxes in the lower right-hand corner are the matrices for the irreducible representation A_1. Note that the traces of these box diagonal matrices give the characters for the E and A_1 representations and the characters for the A_2 irreducible representation can then be found from the orthogonality condition.

The transformation matrices for the three classes of symmetry elements operating on the vector components x,y,z as shown above shows that the z component acts as a basis for the A_1 irreducible representation while the components x and y transform into combinations of each other according to the irreducible representation E. Thus the set (x,y) form the basis for E.

The rotation axis R_z remains unchanged under operations of the E and C_3 classes but it changes sign under an operation of the σ class. Thus it transforms according to the A_2 irreducible representation. The other two rotation axes R_x and R_y transform into combinations of each other and therefore form a basis for the E irreducible representation.

Finally consider how the product of vector components transforms in this group. The conventional way to write the components of an axial vector formed by the product of two vectors is given in (2.17). The way the individual components transform under the symmetry operations of this group was described above, and this information can be used to determine how the product of these components transform. Then transformation matrices can be constructed for each symmetry element, and their traces are calculated to determine the characters of this reducible representation as done previously.

The six-dimensional column matrix for a vector product is

$$\begin{pmatrix} x^2 \\ y^2 \\ z^2 \\ 2yz \\ 2xz \\ 2xy \end{pmatrix}. \qquad\qquad (2.17)$$

Using this as a basis vector, the transformation vectors for an element of each class are written as

$$E = \begin{pmatrix} 1 & 0 & 0 & 0 & 0 & 0 \\ 0 & 1 & 0 & 0 & 0 & 0 \\ 0 & 0 & 1 & 0 & 0 & 0 \\ 0 & 0 & 0 & 1 & 0 & 0 \\ 0 & 0 & 0 & 0 & 1 & 0 \\ 0 & 0 & 0 & 0 & 0 & 1 \end{pmatrix} \qquad \therefore \gamma_E = 6,$$

$$\sigma_1 = \begin{pmatrix} 1 & 0 & 0 & 0 & 0 & 0 \\ 0 & 1 & 0 & 0 & 0 & 0 \\ 0 & 0 & 1 & 0 & 0 & 0 \\ 0 & 0 & 0 & 1 & 0 & 0 \\ 0 & 0 & 0 & 0 & -1 & 0 \\ 0 & 0 & 0 & 0 & 0 & -1 \end{pmatrix} \qquad \therefore \gamma_{\sigma_1} = 2,$$

$$C_3 = \begin{pmatrix} \frac{1}{4} & \frac{3}{4} & 0 & 0 & 0 & \frac{\sqrt{3}}{4} \\ \frac{3}{4} & \frac{1}{4} & 0 & 0 & 0 & \frac{\sqrt{3}}{4} \\ 0 & 0 & 1 & 0 & 0 & 0 \\ 0 & 0 & 0 & -\frac{1}{2} & -\frac{\sqrt{3}}{2} & 0 \\ 0 & 0 & 0 & \frac{\sqrt{3}}{2} & -\frac{1}{2} & 0 \\ \frac{\sqrt{3}}{2} & -\frac{\sqrt{3}}{2} & 0 & 0 & 0 & -\frac{1}{2} \end{pmatrix} \qquad \therefore \gamma_{C_3} = 0$$

This irreducible representation can be reduced in terms of $2E$ and $2A_1$ irreducible representations. By observing the transformation properties, it can be seen that z^2 forms a basis function for one of the A_1 irreducible representations while (x^2+y^2) forms a basis function for the other one. One of the E representations has the set (xz,yz) for a basis function and the other has the basis function set (x^2-y^2,xy).

If all of the information on basis functions is included in the character table for this group given above, it is identical with Table 2.17. This shows that the symmetry group for an equilateral triangle is point group C_{3v}.

For some applications it is important to take the direct product of representations and reduce the results in terms of the irreducible representations of the group. As an example for this group, the direct product of the E representation with itself is found by multiplying the character of the E representation for each symmetry class with itself. This gives the characters 4, 1, and 0 for the E, C_3, and σ classes of symmetry operations, respectively. These are the characters of a reducible representation and (2.10) can be used to show that this reduces to one E, one A_1, and one A_2 irreducible representations.

2.3.2 O_h Point Group

One of the most important symmetries in solid state physics is a regular octahedron with a center of inversion symmetry. This describes seven atoms arrayed along the x, y, and z axes of a cube as shown in Fig. 2.3 with the positions of the ions given in Table 2.34. Each side of the cube has a length $2a$. The angle φ is measured around the z-axis in the xy plane counterclockwise from the x-axis. The angle θ is measured around the y-axis in the xz plane counterclockwise from the z-axis.

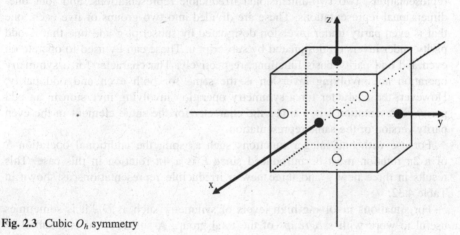

Fig. 2.3 Cubic O_h symmetry

Table 2.34 Ion positions in Fig. 2.3

x	y	z	r	θ	φ
0	0	0	0	0	0
a	0	0	a	$\pi/2$	0
0	a	0	a	$\pi/2$	$\pi/2$
$-a$	0	0	a	$\pi/2$	π
0	$-a$	0	a	$\pi/2$	$3\pi/2$
0	0	a	a	0	0
0	0	$-a$	a	π	0

The cubic O_h point group describes this symmetry. This group has 48 symmetry elements divided into 10 classes. The character table for the group is given in Table 2.32. The symmetry elements are as follows. First is the identity operation E and the inversion operation i which each forms a class by itself. Next there is a class of six C_4 elements describing $\pm 90°$ rotations about the x, y, or z axes. Then there are three twofold axes of rotation C_2 about the x, y, or z axes. Also there are six twofold rotation axes C_2' that run from the center of an edge through the center of the cube to the center of the opposite edge. There are eight axes of $\pm 120°$ rotation about the body diagonals of the cube. There are three mirror planes of symmetry going through the centers of the edges of the cube in the xy, xz, and yz planes. Note that these are equivalent to combined $C_2 i$ operations. Similarly, there are six diagonal planes of symmetry equivalent to combined $C_2' i$ operations. The reflection operations can also be combined with C_3 rotations and C_4 rotations to give eight S_6 and six S_4 operations, respectively.

Since there are 10 classes, there must be 10 irreducible representations for the O_h point group. The only way for the sum of the squares of the dimensions of 10 irreducible representations to equal the order of the group, 48, is $4(3)^2 + 2(2)^2 + 4(1)^2 = 48$. This shows that the group has four three-dimensional irreducible representations, two two-dimensional irreducible representations, and four one-dimensional representations. These are divided into two groups of five each, one that is even parity under inversion designated by subscript g and one that is odd parity under inversion designated by subscript u. These can by used to operate on even and odd parity basis functions, respectively. The character for a symmetry operation not involving inversion is the same for both even and odd parity. However, the character for a symmetry operation involving inversion in an odd parity representation is -1 times the character for the same element in the even parity version of the same representation.

For use with half-integer functions such as spin, the additional operation R of a 2π rotation must be introduced since E is a 4π rotation in this case. This results in three new g and three new u irreducible representations as shown in Table 4.32.

For situations involving high levels of symmetry such as O_h, it is sometimes useful to work with *subgroups* of the total group. A subgroup of is a subset of elements of the larger group that by themselves obey all of the mathematical requirements to be a group. For example, D_{3d} forms a subgroup of O_h consisting of the identity element, two threefold rotation operations, three C_2' operations, and the inversion operation multiplied by each of these elements. The character table for D_{3d} is given in Table 2.30. The irreducible representations of the group can be

O_h	D_{3d}
A_{1g}	A_{1g}
A_{2g}	A_{2g}
E_g	E_g
T_{1g}	$A_{2g} + E_g$
T_{2g}	$A_{1g} + E_g$

decomposed in terms of combinations of the irreducible representations of the subgroup. Comparing the characters of the common elements in O_h and D_{3d} shows the correlation between the irreducible representations of the group and its subgroup. For the even parity representations this is:

The results of this method of inspection can be checked against the predictions of (2.10). For example, for the T_{2g} irreducible representation of O_h the A_{1g} irreducible representation of D_{3d} will appear the following number of times:

$$n^{(T_{2g})} = \frac{1}{12}(1 \times 1 \times 3 + 2 \times 1 \times 0 + 3 \times 1 \times 1 + 1 \times 1 \times 3 + 2 \times 1 \times 0 + 3 \times 1 \times 1)$$
$$= 1,$$

while the A_{2g} irreducible representation will appear the following number of times:

$$n^{(T_{2g})} = \frac{1}{12}(1 \times 1 \times 3 + 2 \times 1 \times 0 + 3 \times (-1) \times 1 + 1 \times 1 \times 3 + 2 \times 1 \times 0 + 3$$
$$\times (-1) \times 1)$$
$$= 0.$$

This is consistent with the correlation table shown above.

From Table 2.32 it can be seen that the vector components (x,y,z) transform as the T_{1u} irreducible representations. As discussed in Sect. 2.4 and in Chap. 4, this is important in using group theory to determine allowed electromagnetic transitions. Also the irreducible representations for O_h involving half-integer quantities are shown in the table. Section 4.4 describes an example of using these representations for atoms with half-integer values of angular momentum.

2.4 Group Theory in Quantum Mechanics

In quantum mechanics a physical system is described by a Hamiltonian operator H. The allowed states of a system are described by a set of orthonormal eigenfunctions ψ_n and the energy of these states is a set of eigenvalues E_n. The sets of eigenfunctions and eigenvalues for the system are found by solving the *Schrödinger equation*

$$H|\psi_n\rangle = E_n|\psi_n\rangle$$

or

$$E_n = \langle\psi_n|H|\psi_n\rangle. \tag{2.18}$$

Since the Hamiltonian describes the physical system, it should be invariant under the same symmetry operations that leave the physical system invariant [2,5]. This is described as a similarity transformation on H by a symmetry operator A

$$H = A^{-1}HA. \tag{2.19}$$

This is the same as saying that a symmetry operator commutes with the Hamiltonian operator [2, 5]. The symmetry operators that leave H invariant form a group. Obviously an operator that does not change H at all is an element of the group and this is the identity operator. For two elements A and B

$$AHA^{-1} = H \quad \text{and} \quad BHB^{-1} = H \quad \text{so} (AB)H(AB)^{-1} = (AB)H(B^{-1}A^{-1}) = H,$$

which shows that the product of two elements is an element that leaves H invariant. Also, the associative law holds. Finally, if (2.19) is multiplied from the left with A and from the right with A^{-1} gives

$$AHA^{-1} = H,$$

which shows that the inverse of the element also leaves H invariant. Thus all the elements that leave H invariant conform to the properties of a group. This is called the group of the Schrödinger equation or the group of the Hamiltonian and is the same as the symmetry group of the system described by H.

If one of the operators of the group of the Hamiltonian A is applied to the initial Schrödinger equation given above,

$$AH|\psi_n\rangle = E_n A|\psi_n\rangle$$

or

$$H|A\psi_n\rangle = E_n|A\psi_n\rangle$$

since A and H commute. This shows that $|A\psi_n\rangle$ is also an eigenfunction belonging to the same eigenvalue E_n. In other words, the eigenfunctions transformed by an operator of the group of H belong to the same eigenvalue as the initial eigenfunctions. From (2.18),

$$E_n = \langle\psi_n|H|\psi_n\rangle = \langle\psi_n|A^{-1}HA|\psi_n\rangle = \langle A^\dagger\psi_n|H|A\psi_n\rangle = \langle\varphi_n|H|\varphi_n\rangle \tag{2.20}$$

where

$$|\varphi_n\rangle = A|\varphi_n\rangle = |A\varphi_n\rangle. \tag{2.21}$$

This derivation uses the fact that for symmetry operators in quantum mechanics their inverse is equal to their adjoint $A^{-1} = A\dagger$.

If E_n has only one eigenfunction then $\varphi_n = \psi_n$ except for a possible phase factor and E_n is said to be *nondegenerate*. If an eigenvalue has associated with it an orthonormal set of eigenfunctions, it is said to be degenerate, and any normalized linear combination of these eigenfunctions will also have the eigenvalue E_n. This is expressed as

$$E_n = \langle \Psi_I | H | \Psi_I \rangle$$

where

$$|\Psi_I\rangle = \sum_i^n a_i |\psi_i\rangle$$

and $|\Psi_I\rangle$ is normalized and $a_1^2 + a_2^2 + \cdots + a_n^2 = 1$. Thus

$$\langle \Psi_I | \Psi_I \rangle = \sum_{i=1}^n |a_i|^2 \langle \psi_i | \psi_i \rangle = 1.$$

There are n possible linear orthogonal combinations.

A symmetry operation of the system acting on a set of degenerate eigenfunctions takes them into a different linear orthogonal combination of the degenerate eigenfunctions:

$$A \begin{pmatrix} |\psi_1\rangle \\ |\psi_2\rangle \\ \vdots \\ |\psi_n\rangle \end{pmatrix} = \begin{pmatrix} a_{11}|\psi_1\rangle + a_{12}|\psi_2\rangle + \cdots + a_{1n}|\psi_n\rangle \\ a_{21}|\psi_1\rangle + a_{22}|\psi_2\rangle + \cdots + a_{2n}|\psi_n\rangle \\ \vdots \\ a_{n1}|\psi_1\rangle + a_{n2}|\psi_2\rangle + \cdots + a_{nn}|\psi_n\rangle \end{pmatrix}. \tag{2.22}$$

The discussion above shows that for quantum mechanical systems, if all symmetry operations for the system leave a specific eigenfunction unchanged (except for a phase factor), that function transforms like a nondegenerate solution of the Schrödinger equation. If some of the symmetry operations act on an eigenfunction to create new linearly independent eigenfunctions, all of these functions transform like members of a degenerate set of solutions to the Schrödinger equation.

From the discussion above, it can be seen that the eigenfunctions belonging to the same eigenvalue of a quantum mechanical system form a basis for one of the irreducible representations of group describing the system. The dimension of the irreducible representation is the same as the degeneracy of the eigenvalue. Thus

$$HA|\psi_i\rangle = E_i A|\psi_i\rangle$$

shows that both $|\psi_i\rangle$ and $A|\psi_i\rangle$ are eigenfunctions of E_i. If E_i is nondegenerate and the eigenfunctions are normalized, $A|\psi_i\rangle = \pm 1|\psi_i\rangle$. Applying all of the symmetry operations of the group generates a one-dimensional irreducible representation of the group with matrix elements (and characters) ± 1. That irreducible representation thus can be used to represent the energy state of the system associated with the eigenvalue for that specific eigenfunction. Considering the same procedure for a degenerate state of the system generates an irreducible representation of the system whose dimension is equal to the degeneracy of the state it represents.

Consider the example of a system with C_{3v} symmetry described in Sect. 2.3.1. A quantum mechanical system with this symmetry will have a nondegenerate eigenfunction that is the basis for the A_1 irreducible representation so it remains unchanged under all symmetry operations. It will have another nondegenerate eigenfunction that remains invariant under operations of the E and C_3 class but changes sign under σ class operations and therefore is the basis for the A_2 irreducible representation. Two other degenerate eigenfunctions will form the basis for the two-dimensional E irreducible representation. If this is designated Γ_3,

$$A(\psi_1 \psi_2) = (\psi_1 \psi_2) \begin{pmatrix} \Gamma_3(A)_{11} \Gamma_3(A)_{12} \\ \Gamma_3(A)_{21} \Gamma_3(A)_{22} \end{pmatrix},$$

where A is a symmetry operator in C_{3v}. This leads to

$$A\psi_1 = \Gamma_3(A)_{11}\psi_1 + \Gamma_3(A)_{21}\psi_2,$$
$$A\psi_2 = \Gamma_3(A)_{12}\psi_1 + \Gamma_3(A)_{22}\psi_2.$$

This shows that ψ_1 transforms like the first column of the transformation matrix of the symmetry operator while ψ_2 transforms like the second column.

When spin–orbit interaction is important, the total wavefunction describing the system is the product of spatial and spin functions:

$$\Psi_i = \psi_i \sigma_i,$$

where σ_i represents the spin angular momentum of the system. In this case double-valued representations must by used. An example of this is given in Chap. 4.

Any physical process interacting with the system can be expressed as a quantum mechanical operator which also transforms according to one of the irreducible representations of the group. The transformation of the specific ith operator O_i^n of a set of n operators is expressed as

$$AO_i^n A^{-1} = \sum_j O_j^n \Gamma_n(A)_{ji}, \tag{2.23}$$

where A is an element of the group of the Hamiltonian and Γ_n is a representation of this group.

The physical process may cause the system to undergo transitions from one eigenstate to another or to split degenerate energy states into several states with lower degeneracies. The qualitative features of these effects can be determined by using group theory techniques such as the direct products and decompositions of representations to evaluate matrix elements. In general the quantum mechanical description of these physical processes involves evaluating matrix elements

$$\langle \psi_f | O | \psi_i \rangle$$

where i and f designate initial and final states of the system and O represents the physical operator. The matrix element represents an integral over all space and to be nonzero the integrand must be symmetric. Instead of evaluating the complete mathematical expression for the integral of the products of the operator and eigenfunctions, we can rewrite this as the product of the group theory representations of the functions

$$\langle \psi_f | O | \psi_i \rangle \neq 0, \quad \text{if } \Gamma_f \times \Gamma_O \times \Gamma_i = A_{1g} + \cdots \qquad (2.24)$$

and

$$\langle \psi_f | O | \psi_i \rangle = 0, \quad \text{if } \Gamma_f \times \Gamma_O \times \Gamma_i \neq A_{1g} + \cdots.$$

Thus the matrix element is nonzero if the decomposition of the triple direct product representation $\Gamma_f \times \Gamma_o \times \Gamma_i$ contains the totally symmetric A_{1g} representation. This is called an *allowed transition*. The matrix element is zero if it does not contain A_{1g}. This is called a *forbidden transition*. This can be stated in a different way knowing that A_{1g} will only appear in the decomposition of the direct product of a representation with itself. Thus for a nonzero matrix element the decomposition of the direct product representation of the initial and final states must contain the irreducible representation of the operator causing the transition.

These concepts can be visualized using a simple example of rectangular symmetry. If the square symmetry shown in Fig. 1.2 is stretched along the x direction the symmetry group is lowered from D_{4h} to D_{2h} with the character table given in Table 2.8. A quantum mechanical state of the system is designated by one of the eight irreducible representations listed in the character table. The signs of the eigenfunctions transforming as some of these representations within the rectangular space are shown in Fig. 2.4. The effect of a symmetry operation on the sign of the function is given by the character of the operation. A positive character leaves the sign unchanged while a negative character changes the sign. For example, the function transforming as the totally symmetric irreducible representation A_g is positive throughout the rectangular space and does not change when it undergoes under any of the symmetry operations. The function transforming as B_{1g} changes sign under the $C_2(y)$, $C_2(x)$, $\sigma(xz)$, and $\sigma(yz)$ operations, all of which have characters of -1, and remains unchanged under the other four operations which have characters of $+1$. The function transforming as B_{2g} changes sign under the $C_2(y)$, $C_2(z)$,

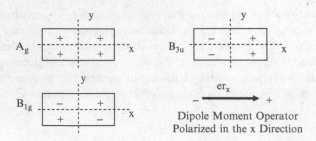

Fig. 2.4 Signs of the basis functions for some of the irreducible representations in a system with rectangular symmetry

ι, and $\sigma(yz)$ operations. The triple product of the three irreducible representations shown in Fig. 2.4 is a function that has positive values in the two upper quadrants and negative values in the two lower quadrants. This forms the basis of a B_{2u} representation. Integrating this function over the area of the rectangle is identically zero since there are equal positive and negative areas. The dipole moment operator polarized in the x direction is also shown in Fig. 2.4. Applying the symmetry operations of the D_{2h} point group shows that this transforms according to the B_{3u} representation as indicated in the character table for the group. The fact that the reduction of the triple direct product $A_g \times B_{3u} \times B_{1g} = B_{2u}$ does not contain A_g is consistent with the fact that the matrix element is zero and the electric dipole induced transition between states A_g and B_{1g} is forbidden. The determination of transition matrix elements in this way is discussed further in later chapters.

Using these concepts of group theory, the irreducible representations of the group of symmetry operations that leave the Hamiltonian of the quantum mechanical system invariant provide information about the degrees of degeneracy of the eigenfunctions of the system and the transformation properties of these eigenfunctions. The group theory procedure of forming and decomposing direct products of representations is useful in quickly determining qualitatively whether a transition is allowed or forbidden, or how many states occur in the splitting of an energy level. However, it can not provide quantitative information about these processes.

2.5 Problems

Consider the thin, square object with the basis set shown in the figure. (○ represents objects above the plane of the square and • represents objects below this plane.) The z-axis is directed out of the paper from the center of the square. Answer the following questions:

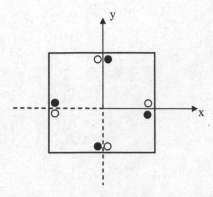

1. Identify the symmetry elements for the point group of this object.
2. Develop the multiplication table for the symmetry elements.
3. Derive the classes of elements for this symmetry group. What is the order of the group?
4. Derive the transition matrices for the group elements operating on the x,y,z coordinates and find the character of each of these.
5. Derive the character table for this group using the concepts of box diagonalization and the properties of characters (especially (2.11) and (1.12)).

References

1. M.S. Dresselhaus, G. Dresselhaus, A. Jorio, *Group Theory* (Springer, Berlin, 2008)
2. M. Tinkham, *Group Theory and Quantum Mechanics* (McGraw-Hill, New York, 1964)
3. M. Lax, *Symmetry Principles in Solid State and Molecular Physics* (Wiley, New York, 1974)
4. F.A. Cotton, *Chemical Applications of Group Theory* (Wiley, New York, 1963)
5. J.L. Prather, *Atomic Energy Levels in Crystals* (Department of Commerce, Washington, 1961)
6. B. DiBartolo, R.C. Powell, *Phonons and Resonances in Solids* (Wiley, New York, 1976)
7. R.S. Knox, A. Gold, *Symmetry in the Solid State* (Benjamin, New York, 1964)
8. M. Hamermesh, *Group Theory and Its Application to Physical Problems* (Addison-Wesley, Reading, 1962)
9. P.T. Landsberg, *Solid State Theory* (Wiley, New York, 1969)

Chapter 3
Tensor Properties of Crystals

Due to the spatial symmetry of crystals discussed in Chap. 1, many important physical properties of a crystalline material depend on the orientation of the sample with respect to some specifically defined coordinate directions. Examples of such properties are electrical conductivity, elasticity, the piezoelectric effect, and nonlinear optics. The last of these is treated in detail in Chap. 6. In this chapter the tensor operator formalism that uses symmetry for treating these properties is summarized, and some specific examples are presented. The discussion focuses on bulk, macroscopic properties of materials and does not include surface effects or microscopic phenomena. The concepts of group theory developed in Chap. 2 are useful in elucidating the qualitative characteristics of these physical properties in materials with different crystal structures. The quantitative values for these properties must be found by experimental measurement. The tensor operator approach to material properties summarized here was developed extensively by Nye and is presented in detail in his book on the topic [1]. It should be noted that not all physical properties can be represented as tensors. Examples of nontensor properties include dielectric strength and surface hardness.

The fundamental question of interest in this analysis [1,2] is, "What is the physical effect that a specific physical cause will have on a crystal with a specific structure?" The physical property of the material under study relates the effect to the cause. Note that the cause under consideration might be external to the crystal, such as an applied electric field, or internal, such as a local electric field supported by the geometry of the crystal structure. When the physical cause with specific directional properties is applied to the crystal, group theory can be used to determine the directional properties of the effect produced by this cause. This basic concept is demonstrated by the fundamental equation

$$\bar{\bar{E}} = \bar{\bar{M}}\bar{\bar{C}} \tag{3.1}$$

Here $\bar{\bar{C}}$ is a tensor representing the physical cause of an effect represented by another tensor $\bar{\bar{E}}$. These are called *field tensors*. The *matter tensor* $\bar{\bar{M}}$ represents the physical property of the crystal that relates the cause and effect. According to a fundamental postulate of crystal physics known as Neumann's Principle,

R.C. Powell, *Symmetry, Group Theory, and the Physical Properties of Crystals*,
Lecture Notes in Physics 824, DOI 10.1007/978-1-4419-7598-0_3,
© Springer Science+Business Media, LLC 2010

the symmetry of a physical property of a crystal must include the same spatial symmetry characteristics as the crystal structure and thus the symmetry of the matter tensor must include all of the symmetry operations contained in the point group of the crystal. This principle holds under the conditions of interest for this chapter as stated in the previous paragraph. Note that nonsymmorphic space groups may involve small translations on the atomic scale coupled with rotational operations. For the macroscopic properties of interest here, such small translations will not make a major contribution to the properties being measured, and thus only the symmetry operations of the crystallographic point group are important.

As an example, (3.1) can be expressed as a Taylor series expansion

$$E_i(C_j + dC_j) = E_i(C_j) + \left(\frac{\partial E_i}{\partial C_j}\right)dC_j + \frac{1}{2}\left(\frac{\partial^2 E_i}{\partial C_j \partial C_k}\right)dC_j dC_k - \cdots.$$

Here the field vectors $\vec{C}$ and $\vec{E}$ represent the cause and effect vectors while the nine quantities $(\partial E_i / \partial C_j)$ are the components of a second-rank matter tensor and the 27 quantities $(\partial^2 E_i / \partial C_j \partial C_k)$ are the components of a third-rank matter tensor. The higher terms in the expansion give rise to matter tensors of higher rank. This type of tensor analysis approach is discussed below for physical properties described by tensors of different ranks.

The statement of Neumann's Principle above does not require that the symmetry group of the matter tensor be exactly the same as the point group of the crystal. It must include all of the symmetry elements of the crystallographic point but can also include additional symmetry elements not in the point group of the crystal.

Physical properties having no directional characteristics are described by scalars regardless of the symmetry of the crystal. Temperature is an example of this. Scalars can be represented as tensors of rank zero, and for induced properties the matter tensor is expressed as a scalar times the unit tensor. In addition, for isotropic materials there are no special degrees of symmetry so there are no intrinsic directional characteristics of any physical property in these materials. Since we have no symmetry operations to use, group theory is not a useful tool for scalar properties or isotropic materials.

A matter tensor of first rank (vector) relates a physical cause represented by a scalar to an effect represented by a vector. For cause and effect phenomena that are both described by vectors, the matter tensor is a second-rank tensor given by a 3×3 matrix. Other physical phenomena may be represented by a matter tensor of third rank that relates a cause tensor of second rank to an effect tensor of first rank. A matter tensor of fourth rank can be used to relate cause and effect tensors of second rank. In each of these cases the symmetry properties of the crystal determine which of the individual elements of the matter tensor are nonzero and which ones must be equal to one another. Note that symmetry properties alone cannot determine the magnitudes of the matter tensor elements. This must be left to experimental measurement.

3.1 First-Rank Matter Tensors

Matter tensors of first rank connect a scalar cause with a vector effect. This can be thought of as having two causes, the scalar cause which is isotropic, and a cause associated with an intrinsic property of the crystal, which is anisotropic. An example of this type of property is the *pyroelectric effect* that relates a change in temperature (scalar) to the induced electric polarization (vector). This occurs because the centers of gravity of the positive and negative charges are separated in their positions in the unit cell of the crystal. As the temperature changes, this distance changes resulting in a change in the electrical polarization. The asymmetry of the combined causes must be present before the change in temperature occurs. The combined effect of the two causes must have the symmetry of the crystal. As shown below, this limits the types of crystal symmetries that exhibit this type of physical effect.

The pyroelectric effect is expressed as

$$\mathbf{P} = \mathbf{p}\Delta T \tag{3.2}$$

or in component form as

$$P_i = p_i \Delta T. \tag{3.3}$$

Here $\mathbf{P}$ is the temperature-induced electric polarization vector with components P_i, and $\mathbf{p}$ is the pyroelectric matter tensor with components p_i. When a material is heated, thermal expansion occurs. On the atomic scale, if the electric charge distribution is not uniform in all directions, this can result in local electric dipole moments. If these dipoles are aligned in the same direction throughout the crystal, they result in a macroscopic electric polarization of the material.

According to Neumann's Principle, $\mathbf{p}$ must be invariant under all operations of the point group of the crystal. The effect of a spatial rotation of the crystal on a first-rank tensor is described by the process developed in Chap. 2. For rotation represented by R,

$$\mathbf{p}' = R\mathbf{p} \tag{3.4}$$

or in component form as

$$p_j' = \sum_{i=1}^{3} R_{ji} p_i. \tag{3.5}$$

Thus if rotation R is part of the crystallographic point group, $\mathbf{p}'$ must equal $\mathbf{p}$.

As an example of a specific symmetry operation, consider the effect of a C_3 rotational operation on the vector $\mathbf{p}$. From Sect. 2.3.1,

$$\begin{pmatrix} p'_x \\ p'_y \\ p'_z \end{pmatrix} = \begin{pmatrix} -1/2 & \sqrt{3}/2 & 0 \\ -\sqrt{3}/2 & -1/2 & 0 \\ 0 & 0 & 1 \end{pmatrix} \begin{pmatrix} p_x \\ p_y \\ p_z \end{pmatrix}$$

so $p'_x = -(1/2)p_x + (\sqrt{3}/2)p_y$, $p'_y = -(\sqrt{3}/2)p_x - (1/2)p_y$, and $p'_z = p_z$. The only way for $\mathbf{p}$ to be invariant in a crystal with a point group that contains the symmetry operation C_3 is for its components p_x and p_y to be identically zero. Then the pyroelectric vector is oriented along the axis of rotation. Thus crystallographic point groups containing a C_3 operation may exhibit a pyroelectric effect depending on the other symmetry operations that are part of the group.

Next consider the effect of an inversion operation $R(i)$. This symmetry transformation takes p_x into $-p_x$, p_y into $-p_y$, and p_z into $-p_z$. Thus an inversion operation takes $\mathbf{p}$ into $-\mathbf{p}$ without leaving any of its components invariant. This means that all crystals having inversion symmetry cannot exhibit the pyroelectric effect.

This symmetry analysis can be generalized to all point groups without considering each individual symmetry transformation. From the above example, it is apparent that a component of the field vector transforms into itself if part of the transformation matrices operating on this component is the identity representation. The rotation matrix shown for the C_3 operation has a 2×2 and a 1×1 box diagonal matrix on its diagonal leading to one field vector component being nonzero. Equation (2.10) derived in Chap. 2 can then be rewritten to give the number of times the identity representation is contained in the box diagonals of the symmetry operation matrices:

$$N = \frac{1}{h} \sum_R \chi(R), \tag{3.6}$$

where the characters of the identity representation are 1 for all operations. The components of the field vectors act a set of basis functions for certain irreducible representations of the symmetry group of the crystal. It is the characters of these representations that are used in (3.6). For a first-rank tensor (vector), the number of times the identity representation appears on the diagonal is the same as the number of independent components of the basis tensor. The way vector components transform as basis functions is given in the character tables for all of the crystallographic point groups in Tables 2.1–2.32.

If we have a crystal with point group symmetry C_{3v}, the pyroelectric tensor must be invariant under the symmetry operations E, $2C_3$, and $3\sigma_v$. The transformation matrices for a vector basis function have the form shown below which have the characters $\chi(E)=3$, $\chi(C_3)=0$, and $\chi(\sigma_v)=1$ that define an reducible representation that can be reduced using (2.10) into one two-dimensional E irreducible representation and one one-dimensional A_1 irreducible representation of the C_{3v} point group using the characters in Table 2.17. These results show that the z component of a vector basis function,

$$E = \begin{pmatrix} 1 & 0 & 0 \\ 0 & 1 & 0 \\ 0 & 0 & 1 \end{pmatrix}, C_3 = \begin{pmatrix} -\frac{1}{2} & \frac{\sqrt{3}}{2} & 0 \\ -\frac{\sqrt{3}}{2} & -\frac{1}{2} & 0 \\ 0 & 0 & 1 \end{pmatrix}, \sigma_y = \begin{pmatrix} -1 & 0 & 0 \\ 0 & 1 & 0 \\ 0 & 0 & 1 \end{pmatrix},$$

transforms as the identity representation A_1 while the y and z components transform together as the E representation. This can be confirmed by applying (3.6) to the first-rank pyroelectric tensor in a crystal with a C_{3v} point group which gives

$$N_p = \frac{1}{6}[3 + 0 + 0 + 1 + 1 + 1] = 1.$$

Here the characters are the sum of the characters of the A_1 and E irreducible representations according to which the vector components x, y, and z transform. This is consistent with the result obtained above with the p_z component of **p** being the only one that is nonzero.

If we are dealing with a crystal having a point group symmetry of C_i the relevant symmetry operations are E and i that have characters 3 and -3 for transformation matrices having a vector basis function. Equation (3.6) then gives

$$N_p = \frac{1}{2}[3 - 3] = 0$$

showing that crystals having point group symmetry C_i cannot exhibit a pyroelectric effect.

The same analysis applied to all of the crystallographic point groups shows that only crystals having a single rotational axis of symmetry, no center of inversion, and no mirror planes perpendicular to the rotational axis can have pyroelectric properties. These include all 10 of the C_n and C_{nv} point groups whose character tables are given in Tables 2.1–2.32. A summary of the forms of the first-rank matter tensors is given in Table 3.1.

Since both the matter tensor and the effect tensor are first-rank tensors, these vectors must be pointed in the same direction and obey the same symmetry transformation properties. Thus the analysis described above could have just as well been applied to the induced polarization tensor. Some materials that posses an

Table 3.1 Form of first-rank tensors for the crystallographic point groups

C_1	C_s	C_2,C_{2v}	C_4,C_{4v} C_3,C_{3v} C_6,C_{6v}	$C_i,C_{2h},D_2,D_{2h}, S_4,C_{4h},$ $D_4,D_{2d}, D_{4h},S_6,D_{3d},D_3,$ $D_{3h},C_{6h},D_{6h},D_6,D_{3h},T,$ T_d,T_h,O,O_h
$\begin{pmatrix} x \\ y \\ z \end{pmatrix}$	$\begin{pmatrix} x \\ 0 \\ z \end{pmatrix}$	$\begin{pmatrix} 0 \\ y \\ 0 \end{pmatrix}$	$\begin{pmatrix} 0 \\ 0 \\ z \end{pmatrix}$	$\begin{pmatrix} 0 \\ 0 \\ 0 \end{pmatrix}$

intrinsic electric polarization give rise to a *ferroelectric effect*. The spatial distribution of electric charge on the atomic scale produces a local electric dipole moment, and when these microscopic dipole moments are aligned in the same direction throughout the crystal, they produce a macroscopic electric polarization. If the direction of the aligned dipoles and thus the polarization can be reversed by the application of an external electric field, the material is said to exhibit a ferroelectric effect. The analysis given above for the pyroelectric effect holds for this case also and the same ten crystal classes that support the pyroelectric effect may exhibit a ferroelectric effect.

The difference between pyroelectric phenomena and ferroelectric phenomena is the effect of an external electric field. For ferroelectric materials, this changes the structural orientation of the electric dipole moments on the atomic scale. Many ferroelectric crystals undergo a phase transition from one crystal structure to another at some specific temperature. Thus it is possible for a crystal to be ferroelectric at low temperatures but not at high temperatures. In general, most ferroelectric crystals have high temperature, nonferroelectric structures belonging to the D_2, D_{2d} or O_h classes [1]. Below some phase-transition temperature their structures change to one of the C_n or C_{nv} subgroups of these classes that permit the ferroelectric effect. Table 3.2 shows the correlation between these typical nonferroelectric crystal classes and their possible ferroelectric subgroups.

A typical example [3] of this type of crystal is $BaTiO_3$. Above 120°C it has a cubic perovskite structure with O_h symmetry and is not ferroelectric. The barium ion is located at the center of a cube with titanium ions at the cube corners and oxygen ions in the centers of the cube faces. Below 120°C the oxygen octahedron slightly distorts to produce a tetragonal structure with C_{4v} symmetry. The crystal is then ferroelectric with the polar axis in the (0,0,1) direction. At a temperature close to 5°C, the crystal further distorts to give an orthorhombic structure with C_{2v} symmetry. The crystal is still ferroelectric in this phase but the polar axis is now in the (0,1,1) direction. As temperature is further lowered to below −90°C, the structure changes to trigonal with C_{3v} symmetry. For this phase, barium titanate is ferroelectric with the polar axis in the (1,1,1) direction.

Any intrinsic physical property of a crystal described by a vector will be subject to the same symmetry conditions as the pyroelectric and ferroelectric examples given above. Some of these other properties are the heat of polarization, the *electrocaloric effect*, polarization by hydrostatic pressure (*piezoelectric effect*), and an electric field due to a change in temperature.

Table 3.2 Correlation between typical nonferroelectric crystal classes and their ferroelectric subgroups [3]

Nonferroelectric crystal class	Subgroups of ferroelectric crystal classes
D_2	C_2, C_1
D_{2d}	C_{2v}, C_2, C_s, C_1
O_h	$C_{4v}, C_{3v}, C_{2v}, C_s, C_1$

It is important to note the difference in transformation properties of a vector and an axial vector, or pseudovector [4]. An axial vector is the result of a vector cross product between two vectors. Whereas true vectors are odd parity functions, axial vectors are even parity functions. Therefore vectors change sign under inversion operations while pseudovectors do not. For a given symmetry operation, the character for a vector basis function and the character for an axial vector basis function are the same if the operation does not involve inversion while they have opposite signs if the operation does involve inversion. The axial vector basis functions are the same as the rotational axis basis functions in the character tables given in Chap. 2. This can be understood by remembering that the vector cross product involves not only the product of the magnitudes of two vectors but also the direction of the resulting vector is found from the right-hand rule of rotating the first vector toward the second vector. A specific symmetry operation can change this direction of rotation in the same way it changes the direction of a rotational axis.

An example of a physical phenomenon represented by an axial vector is the *ferromagnetic effect* where μ is the axial vector representing the intrinsic magnetic dipole moment of an atom in the crystal. This involves the angular momentum of the electrons

$$\mu = (q/2)\mathbf{r} \times \mathbf{v},$$

where the contribution from electron spin has been omitted for simplicity. The vector cross product of position and velocity vectors results in an axial vector. The symmetry transformation properties of μ can be found by first considering how the $\mathbf{r}$ and $\mathbf{v}$ vector components transform and then the cross product components are formed. For example, an inversion operation takes $\mathbf{r}$ into $-\mathbf{r}$ and $\mathbf{v}$ into $-\mathbf{v}$ but the rotation direction of the vector product does not change so the resultant vector is the same as the initial vector. Thus μ is invariant under inversion.

Other symmetry elements can be tested in the same way, and (3.6) with the rotational axes as basis functions can be used to determine which crystallographic point groups will leave μ invariant and which will not. From these considerations, it can be seen that all of the uniaxial groups such as C_n or D_n plus T and O groups leave μ invariant while any group with σ_v or σ_d mirror planes do not. These results are summarized in Table 3.3. Note that the C_{nv} groups that support vector properties are not allowed for axial vectors (see Table 3.1). On the other hand, groups involving inversion operations and σ_h mirror planes that do not leave a vector invariant are

Table 3.3 Crystallographic point groups that can leave an axial vector invariant

Invariant	Not invariant
C_1, C_s, C_i, C_2,	C_{2v}, C_{3v}, C_{4v}, C_{6v},
C_{2h}, D_2, D_{2h}, C_4,	D_{3h}, D_{4h}, D_{2d}, D_{3d},
D_4, C_{4h}, S_4, C_3,	D_{6h}, T_d, O_h
C_{3h}, D_3, C_6, D_6,	
S_6, C_{6h}, T, T_h, O	

allowed for axial vectors. Therefore crystals with point group symmetries such as C_i, C_{nh} and S_n can exhibit a ferromagnetic effect but not a ferroelectric effect.

To illustrate this, consider the case of C_{3v} symmetry considered above for a vector quantity. For an axial vector (3.6) with the appropriate characters from the A_2 and E representations from Table 2.17 gives

$$N_m = \frac{1}{6}[3 + 0 + 0 - 1 - 1 - 1] = 0$$

which was expected because of the σ_v symmetry planes. Note that the C_3 group which has the same first three symmetry elements as C_{3v} but not the three reflection planes does allow the ferromagnetic effect. If we apply this analysis to the point group C_{2h} which is not allowed for vector phenomena like the ferroelectric effect we find for an axial vector that (see Table 2.5)

$$N_m = \frac{1}{4}[3 - 1 + 3 - 1] = 1.$$

Therefore one of the components of the $\boldsymbol{\mu}$ axial vector will be invariant to the symmetry operations of the C_{2h} group. For this group the ferromagnetic effect is allowed while the vector ferroelectric effect is not.

A thorough treatment of symmetry and magnetic properties of solids involves the considerations of different types of magnetism, domain structure, time reversal operations, and other issues that are beyond the scope of this book. An excellent introduction to these topics can be found in [5].

3.2 Second-Rank Matter Tensors

Matter tensors of second rank can connect two first-rank field tensors, or a scalar and a second-rank tensor. Both of these types of situations are discussed below beginning with the former.

For physical phenomena described by vectors, the cause and effect are represented by first-rank tensors and the matter tensor is a second-rank tensor given by a 3×3 matrix. In this case, (3.1) can be expressed in terms of the vector and matrix elements as

$$\varepsilon_i = \sum_j m_{ij} c_j. \tag{3.7}$$

One way to represent the physical properties described by second-rank tensors is by the geometrical shapes their components generate. These are quadratics such as ellipsoids or hyperboloids. Properties where this can be useful include thermal expansion, thermal conductivity, and the refractive index. Examples of quadratics are given later in this chapter and also in Chaps. 5 and 6.

As an example of a second-rank matter tensor, let us consider the property of electrical conductivity $\overset{\leftrightarrow}{\sigma}$ which relates a current density $\boldsymbol{J}$ to the electric field $\mathbf{E}$ which causes it. The crystal is oriented with respect to a laboratory coordinate system. The current density is proportional to the electric field but its direction in this coordinate system may be different. The equation describing this case is

$$\boldsymbol{J} = \overset{\leftrightarrow}{\sigma}\mathbf{E} \tag{3.8}$$

or

$$\begin{pmatrix} j_x \\ j_y \\ j_z \end{pmatrix} = \begin{pmatrix} \sigma_{xx} & \sigma_{xy} & \sigma_{xz} \\ \sigma_{yx} & \sigma_{yy} & \sigma_{yz} \\ \sigma_{zx} & \sigma_{zy} & \sigma_{zz} \end{pmatrix} \begin{pmatrix} E_x \\ E_y \\ E_z \end{pmatrix}. \tag{3.9}$$

In component form this becomes

$$j_i = \sum_j \sigma_{ij} E_j. \tag{3.10}$$

Now consider the matrix elements of the conductivity matter tensor. This tensor must be invariant under all symmetry operations that leave the crystal invariant, i.e., the crystallographic point group. The effect of a symmetry operation R on a crystal rotates the orientations of the electric field, the current density, and the matter tensor in the laboratory reference frame:

$$\mathbf{E}' = R\mathbf{E}, \quad \boldsymbol{J}' = R\boldsymbol{J}, \quad \text{and} \quad \boldsymbol{J}' = \overset{\leftrightarrow}{\sigma}'\mathbf{E}'.$$

Substituting the first two expressions above into the third one and comparing the result with (3.8) shows that the effect of a rotation on the conductivity tensor is given by

$$\overset{\leftrightarrow}{\sigma}' = R\overset{\leftrightarrow}{\sigma}R^{-1}, \tag{3.11}$$

where R is a matrix representing the symmetry operation and R^{-1} is the inverse of this operation. In component form, this can be written as

$$\sigma'_{ij} = r_{ik}r_{jl}\sigma_{kl}. \tag{3.12}$$

According to Neumann's Principle, if R is a member of the point group of the crystal the matrix elements of $\overset{\leftrightarrow}{\sigma}'$ must be the same as those of $\overset{\leftrightarrow}{\sigma}$.

As an example, consider a crystal with cubic symmetry. The symmetry operations of a cubic point group have been described in Sect. 2.3.1. We can apply these one at a time to the second-rank matter tensor to find its nonzero elements and elements of equivalent magnitude. One of the symmetry operations of the O_h crystallographic point group is a rotation of 90° about the z-axis. Since this takes a vector in the x direction to one in the y direction, a vector in the y direction to one in the $-x$ direction, and leaves a vector in the z direction unchanged, the matrix representing this operation is

$$R(C_{4z}) = \begin{pmatrix} 0 & -1 & 0 \\ 1 & 0 & 0 \\ 0 & 0 & 1 \end{pmatrix}.$$

The inverse of this matrix is

$$R^{-1}(C_{4z}) = \begin{pmatrix} 0 & 1 & 0 \\ -1 & 0 & 0 \\ 0 & 0 & 1 \end{pmatrix}.$$

Using these rotation matrices, (3.11) becomes

$$\vec{\vec{\sigma}}' = \begin{pmatrix} 0 & -1 & 0 \\ 1 & 0 & 0 \\ 0 & 0 & 1 \end{pmatrix} \begin{pmatrix} \sigma_{xx} & \sigma_{xy} & \sigma_{xz} \\ \sigma_{yx} & \sigma_{yy} & \sigma_{yz} \\ \sigma_{zx} & \sigma_{zy} & \sigma_{zz} \end{pmatrix} \begin{pmatrix} 0 & 1 & 0 \\ -1 & 0 & 0 \\ 0 & 0 & 1 \end{pmatrix}.$$

Carrying out the matrix multiplication gives

$$\begin{pmatrix} \sigma'_{xx} & \sigma'_{xy} & \sigma'_{xz} \\ \sigma'_{yx} & \sigma'_{yy} & \sigma'_{yz} \\ \sigma'_{zx} & \sigma'_{zy} & \sigma'_{zz} \end{pmatrix} = \begin{pmatrix} \sigma_{yy} & -\sigma_{yx} & -\sigma_{yz} \\ -\sigma_{xy} & \sigma_{xx} & \sigma_{xz} \\ -\sigma_{zy} & \sigma_{zx} & \sigma_{zz} \end{pmatrix}.$$

But this symmetry operation must leave the matrix elements of the matter tensor unchanged. This will only be true if elements σ_{xz}, σ_{yz}, σ_{zx}, and σ_{zy} are all identically zero. In addition σ_{xx} must equal σ_{yy}, and $\sigma_{xy} = -\sigma_{yx}$. The element σ_{zz} is identically unchanged. Thus the fourfold rotation symmetry dictates the form of the matter tensor to be

$$\vec{\vec{\sigma}} = \begin{pmatrix} \sigma_{xx} & \sigma_{xy} & 0 \\ -\sigma_{xy} & \sigma_{xx} & 0 \\ 0 & 0 & \sigma_{zz} \end{pmatrix}.$$

Similarly, a rotation of 180° about the x-axis takes a vector along the z direction into one in the $-z$ direction, a vector along the y direction into one in $-y$, and leaves a vector in the x direction unchanged. This operation is represented by the matrix

$$R(C_{2z}) = \begin{pmatrix} 1 & 0 & 0 \\ 0 & -1 & 0 \\ 0 & 0 & -1 \end{pmatrix}. \tag{3.13}$$

The inverse of this symmetry operation is represented by exactly the same matrix. Substituting these symmetry matrices into (3.11) along with the matter tensor as modified by the C_{4z} operation gives

$$
\begin{pmatrix} 1 & 0 & 0 \\ 0 & -1 & 0 \\ 0 & 0 & -1 \end{pmatrix} \begin{pmatrix} \sigma_{xx} & \sigma_{xy} & 0 \\ -\sigma_{xy} & \sigma_{xx} & 0 \\ 0 & 0 & \sigma_{zz} \end{pmatrix} \begin{pmatrix} 1 & 0 & 0 \\ 0 & -1 & 0 \\ 0 & 0 & -1 \end{pmatrix} = \begin{pmatrix} \sigma_{xx} & -\sigma_{xy} & 0 \\ \sigma_{xy} & \sigma_{xx} & 0 \\ 0 & 0 & \sigma_{zz} \end{pmatrix}.
$$

Comparing the elements of the matter tensor before and after the rotation shows that the elements σ_{xy} and σ_{yz} are also identically zero giving us a diagonal matter tensor matrix.

Another symmetry operation of the cubic point group is a 90° rotation around the x-axis. This is represented by the matrix

$$
R(C_{4x}) = \begin{pmatrix} 1 & 0 & 0 \\ 0 & 0 & -1 \\ 0 & 1 & 0 \end{pmatrix}
$$

and its inverse

$$
R^{-1}(C_{4x}) = \begin{pmatrix} 1 & 0 & 0 \\ 0 & 0 & 1 \\ 0 & -1 & 0 \end{pmatrix}.
$$

Substituting these rotation operators into (3.11) along with the diagonal matter tensor shows that all three nonzero elements of the conductivity tensor are equal in magnitude. Thus we can express the conductivity tensor as

$$
\bar{\bar{\sigma}} = \sigma \begin{pmatrix} 1 & 0 & 0 \\ 0 & 1 & 0 \\ 0 & 0 & 1 \end{pmatrix}. \tag{3.14}
$$

The remaining operations of the cubic point group do not provide any further simplification. Thus we have simplified the matter tensor as much as it can be for this crystal symmetry through the use of group theory. This shows that for cubic crystals conductivity is isotropic. The magnitude of the scalar σ must be found from experimental measurements.

The components of a second-rank tensor transform like the products of vector components. These are listed as basis functions transforming as specific irreducible representations in the character tables of the 32 crystallographic point groups in Chap. 2. For O_h symmetry there are three combinations of the components of a second-rank tensor that transform as A_{1g}, E_g, and T_{2g} irreducible representations. Using (3.6) and the characters from single group part of O_h in Table 2.32 gives

$$N(A_{1g}) = (1 \times 1 + 8 \times 1 + 6 \times 1 + 6 \times 1 + 3 \times 1 + 1 \times 1 + 6 \times 1 + 8 \times 1 + 3 \times 1 + 6 \times 1)/48 = 1$$

$$N(E_g) = (1 \times 2 + 8 \times (-1) + 6 \times 0 + 6 \times 0 + 3 \times 2 + 1 \times 2 + 6 \times 0 + 8 \times (-1) + 3 \times 2 + 6 \times 0)/48 = 0$$

$$N(T_{2g}) = (1 \times 3 + 8 \times 0 + 6 \times 1 + 6 \times (-1) + 3 \times (-1) + 1 \times 3 + 6 \times (-1) + 8 \times 0 + 3 \times (-1) + 6 \times 1)/48 = 0.$$

This analysis shows that there should be only one unique matrix element for the conductivity tensor in crystals exhibiting O_h symmetry. Thus the prediction of (3.6) is consistent with the derivation of the form of $\overset{=}{\sigma}$ described above.

Table 3.4 gives the form of second-rank matter tensors for different symmetries and shows which of the elements are zero and which elements are equivalent. These are relevant for all properties that can be described by a tensor of second rank. Other examples of second-rank matter tensors connecting two vector phenomena include thermal conductivity, dielectric susceptibility, and magnetic susceptibility. Examples of dielectric and optical properties are given in the later chapters.

As mentioned earlier, a quadratic equation describes a spatial surface that is either an ellipsoid or a hyperboloid. The coefficients of a quadratic equation transform in the same way as the components of a second-rank tensor. For the physical properties represented by a symmetric second-rank tensor, a *representation ellipsoid* can be used to visibly describe the anisotropy of the crystal. The general quadratic equation referenced to a coordinate system x_1, x_2, x_3 can be written as

$$T_{11}x_1^2 + T_{12}x_1x_2 + T_{13}x_1x_3 + T_{21}x_2x_1 + T_{22}x_2^2 + T_{23}x_2x_3 +$$
$$T_{31}x_3x_1 + T_{32}x_3x_2 + T_{33}x_3^2 = 1. \tag{3.15}$$

This can always be rotated in space so that the coordinate axes are pointed along the *principal axes* of the ellipsoid. In this case the quadratic equation simplifies to

$$T_1x_1^2 + T_2x_2^2 + T_3x_3^2 = 1. \tag{3.16}$$

A second-rank tensor referred to its principal axes reduces to a diagonal form

Table 3.4 Form of second-rank tensors for the crystallographic point groups

C_1, C_i	C_2, C_s, C_{2h}	$D_2, C_{2v}D_{2h}$
$\begin{pmatrix} \sigma_{11} & \sigma_{12} & \sigma_{13} \\ \sigma_{12} & \sigma_{22} & \sigma_{23} \\ \sigma_{13} & \sigma_{23} & \sigma_{33} \end{pmatrix}$	$\begin{pmatrix} \sigma_{11} & 0 & \sigma_{13} \\ 0 & \sigma_{22} & 0 \\ \sigma_{13} & 0 & \sigma_{33} \end{pmatrix}$	$\begin{pmatrix} \sigma_{11} & 0 & 0 \\ 0 & \sigma_{22} & 0 \\ 0 & 0 & \sigma_{33} \end{pmatrix}$

$$C_4, S_4, C_{4h}, D_4, C_{4v},$$
$$D_{2d}, D_{4h}, C_3, S_6, D_3,$$
$$C_{3v}, D_{3d}, C_6, C_{3h}, C_{6h}, \qquad\qquad T, T_h, T_d, O, O_h$$
$$D_{6h}, D_6, C_{6v}, D_{3h}$$

$\begin{pmatrix} \sigma_{11} & 0 & 0 \\ 0 & \sigma_{11} & 0 \\ 0 & 0 & \sigma_{33} \end{pmatrix}$	$\begin{pmatrix} \sigma_{11} & 0 & 0 \\ 0 & \sigma_{11} & 0 \\ 0 & 0 & \sigma_{11} \end{pmatrix}$

$$\overset{=}{T} = \begin{pmatrix} T_{11} & T_{12} & T_{31} \\ T_{12} & T_{22} & T_{23} \\ T_{31} & T_{23} & T_{33} \end{pmatrix} \Rightarrow \begin{pmatrix} T_1 & 0 & 0 \\ 0 & T_2 & 0 \\ 0 & 0 & T_3 \end{pmatrix}. \tag{3.17}$$

Comparing the general quadratic equation with the equation for an ellipsoid,

$$\frac{x^2}{A^2} + \frac{y^2}{B^2} + \frac{z^2}{C^2} = 1, \tag{3.18}$$

shows that the values of the tensor components along the directions of the principal axes are $T_1^{-1/2}$ in the x_1 direction, $T_2^{-1/2}$ in the x_2 direction, and $T_3^{-1/2}$ in the x_3 direction. T_1, T_2, and T_3 are known as the principal values of the tensor property. If they all have the same sign, the representation surface is an ellipsoid; if they have different signs, it is a hyperboloid. For triclinic crystal systems the representation surface has no restrictions on its axes. For monoclinic crystal systems the surface has one axis parallel to the twofold axis of the crystal while for orthorhombic systems the axes of the surface are parallel to the crystallographic axes. For trigonal, tetragonal, and hexagonal crystal systems the representation surface has an axis of revolution parallel to the C-axis of the crystal. For cubic crystal classes (and isotropic materials) the representation surface reduces to a spherical shape.

Returning to the example given above for electrical conductivity, using the form of (3.17) for the conductivity tensor in (3.8) shows that if the electric field is directed along one of the principal axes, $\mathbf{J}$ will be parallel to $\mathbf{E}$. The magnitude of $\mathbf{J}$ will be different along each of the principal axes depending on the magnitudes of the principal values of $\overset{=}{\sigma}$. If $\mathbf{E}$ is not along a principal axis $\mathbf{J}$ is not parallel to $\mathbf{E}$.

Second-rank tensors can also represent physical phenomena relating a scalar cause to a second-rank tensor effect [1]. An example of this is *thermal expansion* where a change in temperature induces a strain in the crystal. As discussed above, this situation is equivalent to having a dual cause. The extrinsic change in temperature is isotropic and represented by a tensor of zero rank while the intrinsic property of the crystal is anisotropic and represented by a tensor of second rank. The product of their symmetries must exhibit the same symmetry as the crystal. For small changes in temperature, all components of the strain tensor are proportional to ΔT so

$$\varepsilon_{ij} = \alpha_{ij}\Delta T. \tag{3.19}$$

Here the temperature-induced deformation is described by the strain tensor with components ε_{ij}, and α_{ij} are the components of the thermal expansion tensor. Both of these are second-rank tensors. If the thermal expansion tensor is referred to its principal axes as discussed above, (3.10) becomes

$$\varepsilon_i = \alpha_i\Delta T, \tag{3.20}$$

where $i=1$, 2, or 3 and α_1, α_2, and α_3 are the principal expansion coefficients. The thermally induced strain ellipsoid referenced to the principal axes is expressed as

$$\alpha_1 x_1^2 + \alpha_2 x_2^2 + \alpha_3 x_3^2 = 1. \tag{3.21}$$

The second rank thermal expansion tensor $\vec{\vec{\alpha}}$ must obey Neumann's Principle so the analysis described above for electrical conductivity also applies to this type of matter tensor. Thus the forms of the thermal expansion tensor for the different crystallographic point groups are given in Table 3.4.

3.3 Third-Rank Matter Tensors

Next consider a matter tensor of third rank. This can relate a vector to a second-rank tensor [1]. Examples include the piezoelectric effect and the electrooptical effect. The former effect is discussed below and the latter is discussed in Chap. 5.

In some crystals, an electrical polarization is induced when the crystal is stressed. This property is called the *piezoelectric effect* and a matter tensor of third rank relates the vector polarization that is induced to the second-rank stress tensor that caused it. This is expressed as

$$\mathbf{P} = \vec{\vec{d}}\vec{\sigma} \tag{3.22}$$

or in component form as

$$P_k = \sum_{ij} d_{kij}\sigma_{ij}. \tag{3.23}$$

Here P_k represents the kth vector component of the polarizability induced by the applied stress whose tensor elements are given by σ_{ij}. The d_{kij} are the elements of the piezoelectric matter tensor.

The stress tensor elements are shown in Fig. 3.1. The diagonal elements σ_{ii} represent the normal components of stress while the off-diagonal elements σ_{ij} represent the shear stress components. Tensile stress gives positive values of σ_{ii} while compressive stress gives negative values. For shear stress, $\sigma_{ij}=\sigma_{ji}$ so the stress tensor is symmetric about the diagonal [1].

For the special cases of uniaxial stress and hydrostatic pressure the stress tensor has the forms

$$\vec{\vec{\sigma}}_{\text{uniaxial}} = \begin{pmatrix} \sigma & 0 & 0 \\ 0 & 0 & 0 \\ 0 & 0 & 0 \end{pmatrix} \quad \text{and} \quad \vec{\vec{\sigma}}_{\text{hydrostatic}} = \begin{pmatrix} \sigma & 0 & 0 \\ 0 & \sigma & 0 \\ 0 & 0 & \sigma \end{pmatrix}, \tag{3.24}$$

where the values of σ for the hydrostatic stress tensor are all negative. The special cases of pure shear and simple shear are given by matrices of the form

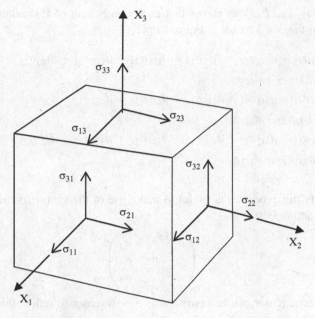

Fig. 3.1 Components of the stress tensor

$$\vec{\vec{\sigma}}_{\text{pure shear}} = \begin{pmatrix} \sigma & 0 & 0 \\ 0 & -\sigma & 0 \\ 0 & 0 & 0 \end{pmatrix} \quad \text{and} \quad \vec{\vec{\sigma}}_{\text{simple shear}} = \begin{pmatrix} 0 & \sigma & 0 \\ \sigma & 0 & 0 \\ 0 & 0 & 0 \end{pmatrix}. \quad (3.25)$$

The piezoelectric tensor $\vec{\vec{d}}$ has the form of a 9×3 matrix divided into three matrices of dimensions 3×3

$$\vec{\vec{d}} = \begin{bmatrix} d_{111} & d_{121} & d_{131} \\ d_{112} & d_{122} & d_{132} \\ d_{113} & d_{123} & d_{133} \\ \hline d_{211} & d_{221} & d_{231} \\ d_{212} & d_{222} & d_{232} \\ d_{213} & d_{223} & d_{233} \\ \hline d_{311} & d_{321} & d_{331} \\ d_{312} & d_{322} & d_{332} \\ d_{313} & d_{323} & d_{333} \end{bmatrix}. \quad (3.26)$$

The rules of multiplication of this tensor in (3.22) or (3.23) is that P_1 is the first row of $\vec{\vec{d}}$ multiplied by the first column of $\vec{\vec{\sigma}}$ plus the second row of $\vec{\vec{d}}$ multiplied by the second column of $\vec{\vec{\sigma}}$ plus the third row of $\vec{\vec{d}}$ multiplied by the third column of $\vec{\vec{\sigma}}$ and the same for P_2 and P_3. This shows that each component of $\mathbf{P}$ is related to every component of the 3×3 tensor $\overset{=}{\sigma}$. For example,

$$
\begin{aligned}
P_1 =&\, d_{111}\sigma_{11} + d_{112}\sigma_{12} + d_{113}\sigma_{13} + d_{121}\sigma_{21} + d_{122}\sigma_{22} + d_{123}\sigma_{23} + d_{131}\sigma_{31} \\
&+ d_{132}\sigma_{32} + d_{133}\sigma_{33}, \\
P_2 =&\, d_{211}\sigma_{11} + d_{212}\sigma_{12} + d_{213}\sigma_{13} + d_{221}\sigma_{21} + d_{222}\sigma_{22} + d_{223}\sigma_{23} + d_{231}\sigma_{31} \\
&+ d_{232}\sigma_{32} + d_{233}\sigma_{33}, \\
P_3 =&\, d_{311}\sigma_{11} + d_{312}\sigma_{12} + d_{313}\sigma_{13} + d_{321}\sigma_{21} + d_{322}\sigma_{22} + d_{323}\sigma_{23} + d_{331}\sigma_{31} \\
&+ d_{332}\sigma_{32} + d_{333}\sigma_{33}.
\end{aligned}
$$

$$(3.27)$$

To simplify this process it is useful to make use of the symmetric nature of the stress tensor and rewrite it as

$$
\overset{=}{\sigma} = \begin{pmatrix} \sigma_1 & \sigma_6 & \sigma_5 \\ & \sigma_2 & \sigma_4 \\ & & \sigma_3 \end{pmatrix}. \tag{3.28}
$$

The piezoelectric tensor can be rewritten as 3×6 tensor to reflect this. Equation (3.22) then can be written as

$$
\begin{pmatrix} P_1 \\ P_2 \\ P_3 \end{pmatrix} = \begin{pmatrix} d_{11} & d_{12} & d_{13} & d_{14} & d_{15} & d_{16} \\ d_{21} & d_{22} & d_{23} & d_{24} & d_{25} & d_{26} \\ d_{31} & d_{32} & d_{33} & d_{34} & d_{35} & d_{36} \end{pmatrix} \begin{pmatrix} \sigma_1 \\ \sigma_2 \\ \sigma_3 \\ \sigma_4 \\ \sigma_5 \\ \sigma_6 \end{pmatrix}. \tag{3.29}
$$

With this construction the matrix multiplication is straightforward, and with the appropriate substitution of indices it gives the same results listed in (3.27).

The symmetry transformation properties of a third-rank tensor such as $\overset{=}{d}$ will determine which of its components are nonzero. These properties can be derived in the same way as was done Sect. 3.2 for a second-rank tensor. Following the rules of transformation for the components of a first-rank tensor, $p_i' = r_{ij}p_j$, and a second-rank tensor, $\sigma_{ij}' = r_{ik}r_{jl}\sigma_{kl}$, the components of a third-rank tensor given in (3.26) transform as

$$
d_{ijk}' = r_{il}r_{jm}r_{kn}d_{lmn}, \tag{3.30}
$$

where the r_{ij} are components of a transformation matrix for a specific symmetry operation. According to Neumann's Principle, the matter tensor components d_{ijk} must remain invariant under the symmetry group operations of the crystal. In this case $d_{ijk}' = d_{lmn}$. For this to be true, the subscripts $l = i$, $m = j$, and $n = k$. Thus only the three diagonal components of the transformation matrix r_{ii}, r_{jj}, and r_{kk} need to be considered. The product of these three elements must be $+1$ for d_{ijk} element of the third-rank tensor to be nonzero.

For example, consider crystals with symmetry groups that contain the inversion operation. Each of the diagonal elements of the matrix representing the inversion operation is equal to -1 so the product of the three of them is also -1. Thus for all components of the piezoelectric tensor, inversion symmetry requires that

$$d'_{ijk} = -d_{ijk}.$$

This can only be true if all of the components are identically zero. Therefore crystals belonging to a symmetry group with a center of inversion do not exhibit a piezoelectric effect.

The same will hold true for groups with a horizontal reflection plane perpendicular to a rotation axis. As an example, consider a rotational axis in the z direction with a horizontal reflection plane σ_h. The matrix for the mirror reflection is

$$\sigma_z = \begin{pmatrix} 1 & 0 & 0 \\ 0 & 1 & 0 \\ 0 & 0 & -1 \end{pmatrix}.$$

The product of the diagonal elements is -1 and all the d_{ijk} elements are identically zero.

This analysis can be used to determine the form of the piezoelectric tensor for the 32 crystallographic symmetry groups and the results are shown in Table 3.5. For a crystal with no spatial symmetry (C_1) all components of $\overset{=}{d}$ may be nonzero. If the only symmetry operation is a horizontal symmetry plane (C_s) then the transformation matrix will have one -1 matrix element on its diagonal and two $+1$ elements (as shown in (2.6)). Thus the components of $\overset{=}{d}$ that are zero will be those that have one or three of the -1 component in their subscripts. For a twofold rotation axis, two of the diagonal elements of the transformation matrix will be -1 and one will be $+1$ (as shown in (2.5)). The nonzero components of $\overset{=}{d}$ will be those having one or three of its subscripts related to the $+1$ element. For example, if the rotation is about the 2-axis, $r_{22}=1$ while $r_{11}=r_{33}=-1$. Thus d_{ij2} and d_{222} elements are nonzero. This is shown for the C_2 class in Table 3.5. If a vertical reflection plane of symmetry containing two of the three major coordinate axes is part of the symmetry group, the transformation matrix will have one diagonal component of -1 and the other two $+1$. Thus any component of $\overset{=}{d}$ having one or three subscripts of the -1 component will be zero. This is reflected in the $\overset{=}{d}$ tensor for the C_{2v} group shown in Table 3.5. This same analysis can be repeated to obtain the form of the piezoelectric tensor for all of the crystal symmetry classes.

In general the applied stress is not isotropic so some of the tensor components of $\overset{=}{\sigma}$ may be zero. As an example, consider a uniaxial stress represented by σ_{11}. If this is applied to a crystal the resulting electrical polarization has the components

$$P_1 = d_{111}\sigma_{11}, \quad P_2 = d_{211}\sigma_{11}, \quad P_3 = d_{311}\sigma_{11}.$$

Then using the components of the piezoelectric tensors given in Table 3.5, it can be seen that for a crystal belonging to symmetry class C_{1h} only the P_1 and P_3

Table 3.5 Form of third-rank tensors for the crystallographic point groups

$$C_i, C_{2h}, D_{2h},$$
$$C_{4h}, D_{4h}, S_6,$$
$$D_{3d}, C_{6h}, D_{6h},$$

C_1
$$\begin{bmatrix} d_{111} & d_{121} & d_{131} \\ d_{112} & d_{122} & d_{132} \\ d_{113} & d_{123} & d_{133} \\ d_{211} & d_{221} & d_{231} \\ d_{212} & d_{222} & d_{232} \\ d_{213} & d_{223} & d_{233} \\ d_{311} & d_{321} & d_{331} \\ d_{312} & d_{322} & d_{332} \\ d_{313} & d_{323} & d_{333} \end{bmatrix}$$

T_h, O, O_h
$$\begin{bmatrix} 0 & 0 & 0 \\ 0 & 0 & 0 \\ 0 & 0 & 0 \\ 0 & 0 & 0 \\ 0 & 0 & 0 \\ 0 & 0 & 0 \\ 0 & 0 & 0 \\ 0 & 0 & 0 \\ 0 & 0 & 0 \end{bmatrix}$$

C_2
$$\begin{bmatrix} 0 & d_{121} & 0 \\ d_{121} & 0 & d_{132} \\ 0 & d_{132} & 0 \\ d_{211} & 0 & d_{231} \\ 0 & d_{222} & 0 \\ d_{231} & 0 & d_{233} \\ 0 & d_{321} & 0 \\ d_{321} & 0 & d_{332} \\ 0 & d_{332} & 0 \end{bmatrix}$$

C_{2v}
$$\begin{bmatrix} 0 & 0 & d_{131} \\ 0 & 0 & 0 \\ d_{131} & 0 & 0 \\ 0 & 0 & 0 \\ 0 & 0 & d_{232} \\ 0 & d_{232} & 0 \\ d_{311} & 0 & 0 \\ 0 & d_{322} & 0 \\ 0 & 0 & d_{333} \end{bmatrix}$$

C_s
$$\begin{bmatrix} d_{111} & 0 & d_{131} \\ 0 & d_{122} & 0 \\ d_{131} & 0 & d_{133} \\ 0 & d_{221} & 0 \\ d_{221} & 0 & d_{232} \\ 0 & d_{232} & 0 \\ d_{311} & 0 & d_{331} \\ 0 & d_{322} & 0 \\ d_{331} & 0 & d_{333} \end{bmatrix}$$

D_2
$$\begin{bmatrix} 0 & 0 & 0 \\ 0 & 0 & d_{132} \\ 0 & d_{132} & 0 \\ 0 & 0 & d_{231} \\ 0 & 0 & 0 \\ d_{231} & 0 & 0 \\ 0 & d_{321} & 0 \\ d_{321} & 0 & 0 \\ 0 & 0 & 0 \end{bmatrix}$$

C_4
$$\begin{bmatrix} 0 & 0 & d_{131} \\ 0 & 0 & d_{132} \\ d_{131} & d_{132} & 0 \\ 0 & 0 & -d_{132} \\ 0 & 0 & d_{131} \\ -d_{132} & d_{131} & 0 \\ d_{311} & 0 & 0 \\ 0 & d_{311} & 0 \\ 0 & 0 & d_{333} \end{bmatrix}$$

D_4
$$\begin{bmatrix} 0 & 0 & 0 \\ 0 & 0 & d_{132} \\ 0 & d_{132} & 0 \\ 0 & 0 & d_{231} \\ 0 & 0 & 0 \\ d_{231} & 0 & 0 \\ 0 & 0 & 0 \\ 0 & 0 & 0 \\ 0 & 0 & 0 \end{bmatrix}$$

S_4
$$\begin{bmatrix} 0 & 0 & d_{131} \\ 0 & 0 & d_{132} \\ d_{131} & d_{132} & 0 \\ -d_{131} & 0 & d_{132} \\ 0 & 0 & 0 \\ d_{132} & 0 & 0 \\ d_{311} & d_{321} & 0 \\ d_{321} & -d_{311} & 0 \\ 0 & 0 & 0 \end{bmatrix}$$

C_{4v}
$$\begin{bmatrix} 0 & 0 & d_{131} \\ 0 & 0 & 0 \\ d_{131} & 0 & 0 \\ 0 & 0 & 0 \\ 0 & 0 & d_{131} \\ 0 & d_{131} & 0 \\ d_{311} & 0 & 0 \\ 0 & d_{311} & 0 \\ 0 & 0 & d_{333} \end{bmatrix}$$

D_{2d}
$$\begin{bmatrix} 0 & 0 & 0 \\ 0 & 0 & d_{132} \\ 0 & d_{132} & 0 \\ 0 & 0 & d_{132} \\ 0 & 0 & 0 \\ d_{132} & 0 & 0 \\ 0 & d_{321} & 0 \\ d_{321} & 0 & 0 \\ 0 & 0 & 0 \end{bmatrix}$$

C_3
$$\begin{bmatrix} d_{111} & 2d_{211} & d_{131} \\ 2d_{211} & -d_{111} & d_{132} \\ d_{131} & d_{132} & 0 \\ d_{211} & -2d_{111} & -d_{132} \\ -2d_{111} & -d_{211} & d_{131} \\ -d_{132} & d_{131} & 0 \\ d_{311} & 0 & 0 \\ 0 & d_{311} & 0 \\ 0 & 0 & d_{333} \end{bmatrix}$$

D_3
$$\begin{bmatrix} d_{111} & 0 & 0 \\ 0 & -d_{111} & d_{132} \\ 0 & d_{132} & 0 \\ 0 & -2d_{111} & -d_{132} \\ -2d_{111} & 0 & 0 \\ -d_{132} & 0 & 0 \\ 0 & 0 & 0 \\ 0 & 0 & 0 \\ 0 & 0 & 0 \end{bmatrix}$$

C_{3v}
$$\begin{bmatrix} 0 & -2d_{222} & d_{131} \\ -2d_{222} & 0 & 0 \\ d_{131} & 0 & 0 \\ -d_{222} & 0 & 0 \\ 0 & d_{222} & d_{131} \\ 0 & d_{131} & 0 \\ d_{311} & 0 & 0 \\ 0 & d_{311} & 0 \\ 0 & 0 & d_{333} \end{bmatrix}$$

C_{3h}
$$\begin{bmatrix} d_{111} & -2d_{222} & 0 \\ -2d_{222} & -d_{111} & 0 \\ 0 & 0 & 0 \\ -d_{222} & 2d_{111} & 0 \\ 2d_{111} & d_{222} & 0 \\ 0 & 0 & 0 \\ 0 & 0 & 0 \\ 0 & 0 & 0 \\ 0 & 0 & 0 \end{bmatrix}$$

D_{3h}
$$\begin{bmatrix} 0 & -2d_{222} & 0 \\ -2d_{222} & 0 & 0 \\ 0 & 0 & 0 \\ -d_{222} & 0 & 0 \\ 0 & d_{222} & 0 \\ 0 & 0 & 0 \\ 0 & 0 & 0 \\ 0 & 0 & 0 \\ 0 & 0 & 0 \end{bmatrix}$$

C_6
$$\begin{bmatrix} 0 & 0 & d_{131} \\ 0 & 0 & d_{132} \\ d_{131} & d_{132} & 0 \\ 0 & 0 & -d_{132} \\ 0 & 0 & d_{131} \\ -d_{132} & d_{131} & 0 \\ d_{311} & 0 & 0 \\ 0 & d_{311} & 0 \\ 0 & 0 & d_{333} \end{bmatrix}$$

C_{6v}
$$\begin{bmatrix} 0 & 0 & d_{232} \\ 0 & 0 & 0 \\ d_{232} & 0 & 0 \\ 0 & 0 & 0 \\ 0 & 0 & d_{232} \\ 0 & d_{232} & 0 \\ d_{311} & 0 & 0 \\ 0 & d_{311} & 0 \\ 0 & 0 & d_{333} \end{bmatrix}$$

D_6
$$\begin{bmatrix} 0 & 0 & 0 \\ 0 & 0 & d_{132} \\ 0 & d_{132} & 0 \\ 0 & 0 & -d_{132} \\ 0 & 0 & 0 \\ -d_{132} & 0 & 0 \\ 0 & 0 & 0 \\ 0 & 0 & 0 \\ 0 & 0 & 0 \end{bmatrix}$$

T, T_d
$$\begin{bmatrix} 0 & 0 & 0 \\ 0 & 0 & d_{132} \\ 0 & d_{132} & 0 \\ 0 & 0 & d_{132} \\ 0 & 0 & 0 \\ d_{132} & 0 & 0 \\ 0 & d_{132} & 0 \\ d_{132} & 0 & 0 \\ d_{132} & 0 & 0 \end{bmatrix}$$

components will be nonzero. On the other hand, for a crystal belonging to class C_2 only the P_2 component will be nonzero and for a crystal with C_{2v} symmetry only the P_3 component will be nonzero.

As a practical example, consider a quartz crystal that has D_3 symmetry at room temperature. The piezoelectric effect for this case is given by

$$
\begin{pmatrix} P_1 \\ P_2 \\ P_3 \end{pmatrix} =
\left[
\begin{array}{ccc|ccc|ccc}
d_{111} & 0 & 0 \\
0 & -d_{111} & d_{132} \\
0 & d_{132} & 0 \\
\hline
0 & -2d_{111} & -d_{132} \\
-2d_{111} & 0 & 0 \\
-d_{132} & 0 & 0 \\
\hline
0 & 0 & 0 \\
0 & 0 & 0 \\
0 & 0 & 0
\end{array}
\right]
\begin{pmatrix} \sigma_{11} & \sigma_{12} & \sigma_{13} \\ \sigma_{21} & \sigma_{22} & \sigma_{23} \\ \sigma_{31} & \sigma_{32} & \sigma_{33} \end{pmatrix},
$$

so

$$P_1 = d_{111}\sigma_{11} - d_{111}\sigma_{22} + d_{132}\sigma_{32} + d_{132}\sigma_{23} = (\sigma_{11} - \sigma_{22})d_{111} + (\sigma_{32} + \sigma_{23})d_{132}$$
$$P_2 = -2d_{111}\sigma_{21} - d_{132}\sigma_{31} - 2d_{111}\sigma_{12} - d_{132}\sigma_{13} = -2d_{111}(\sigma_{21} + \sigma_{12})$$
$$\quad - (\sigma_{13} + \sigma_{31})d_{132}$$
$$P_3 = 0$$

If a uniaxial stress is applied in the σ_{11} direction, $P_1 = d_{11}\sigma_{11}$ and $P_2 = 0$. The same tensile stress applied along σ_{22} also produces a polarization along P_1. The twofold rotation axis P_1 is the electric axis of quartz. Shear stress can produce polarization along P_2 but no stress conditions can produce a polarization along P_3.

The examples given above show the usefulness of symmetry analysis in understanding the third-rank tensor properties of crystals. This is discussed further in Chap. 5 where the electrooptic effect is considered.

3.4 Fourth-Rank Matter Tensors

The tensor analysis of crystal properties described in the preceding sections can be extended to properties represented by fourth-rank tensors. For example, the elastic properties of a crystal are represented by *Hooks's Law*

$$\overset{=}{\varepsilon} = \overset{\equiv}{s}\,\overset{=}{\sigma}. \tag{3.31}$$

Here $\overset{=}{\varepsilon}$ is the second-rank strain tensor, $\overset{=}{\sigma}$ is the second-rank stress tensor, and the fourth-rank matter tensor $\overset{\equiv}{s}$ represents the *elastic compliance*. The extension of (3.23) gives the component form of this expression:

$$\varepsilon_{ij} = \sum_{kl} s_{ijkl}\sigma_{kl}. \tag{3.32}$$

Each of the nine components of $\overset{=}{\varepsilon}$ is linearly related to all nine of the components of $\overset{=}{\sigma}$. The matter tensor $\overset{=}{s}$ is a 9×9 matrix with 81 components. Similar to the discussion in Sect. 3.3, this can be organized as nine different 3×3 matrices.

Because of the complexity of working with an 81-component tensor, a notation has been developed to reduce the problem to a 36-component tensor [1]. This is a 6×6 matrix

$$\overset{=}{s} = \begin{pmatrix} s_{11} & s_{12} & s_{13} & s_{14} & s_{15} & s_{16} \\ s_{21} & s_{22} & s_{23} & s_{24} & s_{25} & s_{26} \\ s_{31} & s_{32} & s_{33} & s_{34} & s_{35} & s_{36} \\ s_{41} & s_{42} & s_{43} & s_{44} & s_{45} & s_{46} \\ s_{51} & s_{52} & s_{53} & s_{54} & s_{55} & s_{56} \\ s_{61} & s_{62} & s_{63} & s_{64} & s_{65} & s_{66} \end{pmatrix}. \tag{3.33}$$

These 36 elements are related to the original 81 elements through the subscript relationships given in Table 3.6.

In the reduced tensor form, (3.32) is rewritten as

$$\varepsilon_i = \sum_j s_{ij}\sigma_j. \tag{3.34}$$

As an example consider the first element of the strain tensor ε_1,

$$\varepsilon_1 = s_{11}\sigma_1 + \frac{1}{2}s_{16}\sigma_6 + \frac{1}{2}s_{15}\sigma_5 + \frac{1}{2}s_{16}\sigma_6 + s_{12}\sigma_2 + \frac{1}{2}s_{14}\sigma_4 + \frac{1}{2}s_{15}\sigma_5 + \frac{1}{2}s_{14}\sigma_4$$
$$+ s_{13}\sigma_3.$$

In expanded subscript form this is

$$\varepsilon_{11} = s_{1111}\sigma_{11} + s_{1112}\sigma_{12} + s_{1113}\sigma_{13} + s_{1121}\sigma_{21} + s_{1122}\sigma_{22} + s_{1123}\sigma_{23} + s_{1131}\sigma_{31}$$
$$+ s_{1132}\sigma_{32} + s_{1133}\sigma_{33}.$$

In using (3.34) in this way, the elements ε_4, ε_5, and ε_6 appear with a factor of ½.

Another way to obtain the correct expansion of the coefficients is to rearrange the elements of $\overset{=}{s}$ to give tensor expression in reduced form as

$$\begin{pmatrix} \varepsilon_1 & \varepsilon_6 & \varepsilon_5 \\ \varepsilon_6 & \varepsilon_2 & \varepsilon_4 \\ \varepsilon_5 & \varepsilon_4 & \varepsilon_3 \end{pmatrix} = \begin{pmatrix} s_{11} & s_{16} & s_{15} & s_{12} & s_{14} & s_{13} \\ s_{21} & s_{26} & s_{25} & s_{22} & s_{24} & s_{23} \\ s_{31} & s_{36} & s_{35} & s_{32} & s_{34} & s_{33} \\ s_{41} & s_{46} & s_{45} & s_{42} & s_{44} & s_{43} \\ s_{51} & s_{56} & s_{55} & s_{52} & s_{54} & s_{53} \\ s_{61} & s_{66} & s_{65} & s_{62} & s_{64} & s_{63} \end{pmatrix} \begin{pmatrix} \sigma_1 & \sigma_6 & \sigma_5 \\ \sigma_6 & \sigma_2 & \sigma_4 \\ \sigma_5 & \sigma_4 & \sigma_3 \end{pmatrix} \tag{3.35}$$

The multiplication rules for obtaining the components of ε in terms of the components of σ are the first three components of the first row of $\overset{=}{s}$ multiply the first
column of σ, the next two components of the first row of $\overset{=}{s}$ multiply the last two
components of the second column of σ, and the last component of the first row of $\overset{=}{s}$
multiplies the last component of the third column of σ. With these multiplication
rules the factors of ½ and ¼ listed in Table 3.6 are not needed.

The same multiplication results can be obtained by treating the six independent
components of the stress tensor as a 1×6 matrix

$$
\begin{pmatrix} \varepsilon_1 \\ \varepsilon_2 \\ \varepsilon_3 \\ \varepsilon_4 \\ \varepsilon_5 \\ \varepsilon_6 \end{pmatrix} = \begin{pmatrix} s_{11} & s_{16} & s_{15} & s_{12} & s_{14} & s_{13} \\ s_{21} & s_{26} & s_{25} & s_{22} & s_{24} & s_{23} \\ s_{31} & s_{36} & s_{35} & s_{32} & s_{34} & s_{33} \\ s_{41} & s_{46} & s_{45} & s_{42} & s_{44} & s_{43} \\ s_{51} & s_{56} & s_{55} & s_{52} & s_{54} & s_{53} \\ s_{61} & s_{66} & s_{65} & s_{62} & s_{64} & s_{63} \end{pmatrix} \begin{pmatrix} \sigma_1 \\ \sigma_6 \\ \sigma_5 \\ \sigma_2 \\ \sigma_4 \\ \sigma_3 \end{pmatrix}. \tag{3.36}
$$

The extension of (3.27) gives the transformation properties of the elastic compliance
components

$$
s'_{ijkl} = r_{im} r_{jn} r_{ko} r_{lp} s_{mnop}, \tag{3.37}
$$

where the r_{ij} are components of a symmetry transformation matrix.

As before, if the transformation operation is a symmetry element of the crystallographic point group, Neumann's Principle requires that the components of the
compliance matter tensor must remain unchanged, $s'_{ijkl} = s_{mnop}$. Equation (3.37)
shows that we need to consider only the diagonal elements of the transformation
matrix, and the product of these four components r_{ii} must be $+1$ to have a nonzero
component of the elastic compliance tensor. Thus, as described in Sect. 3.3, an
inversion operation will have the product of four $r_{ii} = -1$ which is a $+1$ so the
compliance tensor component will be invariant. A similar argument holds for a
σ_h symmetry operation. For operations such as rotation axes or vertical reflection
planes where the diagonal elements of the matrix representing the operation
have some $+1$ and some -1 components, the invariant components of $\overset{=}{s}$ will
be those involving an even number of the -1 components. By applying this
procedure to the symmetry operations of the 32 crystallographic point groups, the
forms of the fourth-rank matter tensor for each class can be determined as was done
above for third-rank tensors. Using the analysis described above with the reduced
tensor notation, the forms of the fourth-rank tensors are given in Table 3.7. These
are in the form of the $\overset{=}{s}$ tensor in (3.33) not the reorganized form of (3.35).

Table 3.6 Relationships between tensor subscripts

Reduced tensor subscripts:	1	2	3	4	5	6
Tensor subscripts:	11	22	33	23,32	31,13	12,21
Numerical factors:	$s_{mn} = s_{ijkl}$ when m and n are 1, 2, 3					
	$s_{mn} = 2s_{ijkl}$ when either m or n is 4, 5, 6					
	$s_{mn} = 4s_{ijkl}$ when both m and n are 4, 5, 6					

Table 3.7 Form of reduced fourth-rank tensors for the crystallographic point groups

C_1, C_i

$$\begin{pmatrix} s_{11} & s_{12} & s_{13} & s_{14} & s_{15} & s_{16} \\ s_{21} & s_{22} & s_{23} & s_{24} & s_{25} & s_{26} \\ s_{31} & s_{32} & s_{33} & s_{34} & s_{35} & s_{36} \\ s_{41} & s_{42} & s_{43} & s_{44} & s_{45} & s_{46} \\ s_{51} & s_{52} & s_{53} & s_{54} & s_{55} & s_{56} \\ s_{61} & s_{62} & s_{63} & s_{64} & s_{65} & s_{66} \end{pmatrix}$$

C_2, C_s, C_{2h}

$$\begin{pmatrix} s_{11} & s_{12} & s_{13} & 0 & s_{15} & 0 \\ s_{21} & s_{22} & s_{23} & 0 & s_{25} & 0 \\ s_{31} & s_{32} & s_{33} & 0 & s_{35} & 0 \\ 0 & 0 & 0 & s_{44} & 0 & s_{46} \\ s_{51} & s_{52} & s_{53} & 0 & s_{55} & 0 \\ 0 & 0 & 0 & s_{64} & 0 & s_{66} \end{pmatrix}$$

C_{2v}, D_2, D_{2h}

$$\begin{pmatrix} s_{11} & s_{12} & s_{13} & 0 & 0 & 0 \\ s_{21} & s_{22} & s_{23} & 0 & 0 & 0 \\ s_{31} & s_{32} & s_{33} & 0 & 0 & 0 \\ 0 & 0 & 0 & s_{44} & 0 & 0 \\ 0 & 0 & 0 & 0 & s_{55} & 0 \\ 0 & 0 & 0 & 0 & 0 & s_{66} \end{pmatrix}$$

$C_4, S_4, C_{4h},$

$$\begin{pmatrix} s_{11} & s_{12} & s_{13} & 0 & 0 & s_{16} \\ s_{21} & s_{11} & s_{13} & 0 & 0 & -s_{16} \\ s_{31} & s_{31} & s_{33} & 0 & 0 & 0 \\ 0 & 0 & 0 & s_{44} & 0 & 0 \\ 0 & 0 & 0 & 0 & s_{44} & 0 \\ s_{61} & -s_{61} & 0 & 0 & 0 & s_{66} \end{pmatrix}$$

$D_4, C_{4v}, D_{2d}, D_{4h}$

$$\begin{pmatrix} s_{11} & s_{12} & s_{13} & 0 & 0 & 0 \\ s_{21} & s_{11} & s_{13} & 0 & 0 & 0 \\ s_{31} & s_{31} & s_{33} & 0 & 0 & 0 \\ 0 & 0 & 0 & s_{44} & 0 & 0 \\ 0 & 0 & 0 & 0 & s_{44} & 0 \\ 0 & 0 & 0 & 0 & 0 & s_{66} \end{pmatrix}$$

C_3, S_6

$$\begin{pmatrix} s_{11} & s_{12} & s_{13} & s_{14} & -s_{25} & 0 \\ s_{21} & s_{11} & s_{13} & -s_{14} & s_{25} & 0 \\ s_{31} & s_{31} & s_{33} & 0 & 0 & 0 \\ s_{41} & -s_{41} & 0 & s_{44} & 0 & 2s_{25} \\ -s_{52} & s_{52} & 0 & 0 & s_{44} & -2s_{14} \\ 0 & 0 & 0 & 2s_{52} & -2s_{41} & 2(s_{11}-s_{12}) \end{pmatrix}$$

D_3, C_{3v}, D_{3d}

$$\begin{pmatrix} s_{11} & s_{12} & s_{13} & s_{14} & 0 & 0 \\ s_{21} & s_{11} & s_{13} & -s_{14} & 0 & 0 \\ s_{31} & s_{31} & s_{33} & 0 & 0 & 0 \\ s_{41} & -s_{41} & 0 & s_{44} & 0 & 0 \\ 0 & 0 & 0 & 0 & s_{44} & -2s_{14} \\ 0 & 0 & 0 & 0 & -2s_{41} & 2(s_{11}-s_{12}) \end{pmatrix}$$

$C_6, C_{3h}, C_{6h}, D_6, C_{6v}, D_{3h}, D_{6h}$

$$\begin{pmatrix} s_{11} & s_{12} & s_{13} & 0 & 0 & 0 \\ s_{21} & s_{11} & s_{13} & 0 & 0 & 0 \\ s_{31} & s_{31} & s_{33} & 0 & 0 & 0 \\ 0 & 0 & 0 & s_{44} & 0 & 0 \\ 0 & 0 & 0 & 0 & s_{44} & 0 \\ 0 & 0 & 0 & 0 & 0 & 2(s_{11}-s_{12}) \end{pmatrix}$$

T, T_h, T_d, O, O_h

$$\begin{pmatrix} s_{11} & s_{12} & s_{12} & 0 & 0 & 0 \\ s_{21} & s_{11} & s_{12} & 0 & 0 & 0 \\ s_{21} & s_{21} & s_{11} & 0 & 0 & 0 \\ 0 & 0 & 0 & s_{44} & 0 & 0 \\ 0 & 0 & 0 & 0 & s_{44} & 0 \\ 0 & 0 & 0 & 0 & 0 & s_{44} \end{pmatrix}$$

As an example, consider a crystal with octahedral symmetry belonging to the O_h point group. Using the form of the $\overset{=}{s}$ for this point group in Table 3.7 shows that there are 12 nonzero components of $\overset{=}{s}$. There are only four independent values and three components have each of these values. Thus using (3.34) or (3.35) gives the components of $\overset{=}{\varepsilon}$ such as

$$
\begin{pmatrix} \varepsilon_1 & & \\ \varepsilon_6 & \varepsilon_2 & \\ \varepsilon_5 & \varepsilon_4 & \varepsilon_3 \end{pmatrix} = \begin{pmatrix} s_{11} & 0 & 0 & s_{12} & 0 & s_{12} \\ s_{21} & 0 & 0 & s_{11} & 0 & s_{12} \\ s_{21} & 0 & 0 & s_{21} & 0 & s_{11} \\ 0 & 0 & 0 & 0 & s_{44} & 0 \\ 0 & 0 & s_{44} & 0 & 0 & 0 \\ 0 & s_{44} & 0 & 0 & 0 & 0 \end{pmatrix} \begin{pmatrix} \sigma_1 & & \\ \sigma_6 & \sigma_2 & \\ \sigma_5 & \sigma_4 & \sigma_3 \end{pmatrix}. \quad (3.38)
$$

Expanding this gives

$$
\begin{aligned}
\varepsilon_1 &= s_{11}\sigma_1 + s_{12}\sigma_2 + s_{12}\sigma_3 = s_{11}\sigma_1 + s_{12}(\sigma_2 + \sigma_3) \\
\varepsilon_2 &= s_{21}\sigma_1 + s_{11}\sigma_2 + s_{12}\sigma_3 \\
\varepsilon_3 &= s_{21}\sigma_1 + s_{21}\sigma_2 + s_{11}\sigma_3 = s_{21}(\sigma_1 + \sigma_2) + s_{11}\sigma_3 \qquad (3.39) \\
\varepsilon_4 &= s_{44}\sigma_4 \\
\varepsilon_5 &= s_{44}\sigma_5 \\
\varepsilon_6 &= s_{44}\sigma_6.
\end{aligned}
$$

This same type of analysis can be used in dealing with other fourth-rank tensors and higher order tensors.

3.5   Problems

1. Consider a crystal with D_{4h} symmetry. Derive the transformation matrices for each of the operations of the group and show how a first-rank tensor transforms under each of these operations. Use (3.6) to show how many components of a first-rank matter tensor are nonzero for a crystal with this symmetry.
2. Use (3.6) to determine how many independent components of a second-rank matter tensor are nonzero for a crystal with D_{4h} symmetry. Derive the form of the second-rank matter tensor for this crystal using the transformation matrices found in problem 1.
3. Use (3.30) to derive the form of a third-rank matter tensor with D_4 symmetry.
4. Derive the form of a fourth-rank matter tensor for a crystal with D_4 symmetry.

References

1. J.F. Nye, *Physical Properties of Crystals, Their Representations by Tensors and Matrices* (Clarendon, Oxford, 1957)
2. S. Bhagavantam, *Crystal Symmetry and Physicsl Properties* (Academic, London, 1966)

3. M.E. Lines, A.M. Glass, *Principles and Applications of Ferroelectrics and Related Materials* (Clarendon, Oxford, 1977)
4. M. Lax, *Symmetry Principles in Solid State and Molecular Physics* (Wiley, New York, 1974)
5. A.S. Borovik-Romanov, H. Grimmer, in *International Tables for Crystallography Volume D, Physical Properties of Crystals*, ed. A. Authier (Kluwer, Dordrecht, 2003), p. 105

Chapter 4
Symmetry Properties of Point Defects in Solids

The properties of solids can be altered by substituting a small concentration of a different type of ion for one of the normal ions of the host lattice. This is referred to as *doping* the solid and is the key to important applications such as solid state laser materials and microelectronics based on n- and p-type doping of semiconductors. The dopant ion can be either at a normal lattice site or at an interstitial site. It acts like a point defect in the host material and its properties are determined largely by its interaction with its nearest-neighbor ions. The maximum possible concentration and the uniformity of the distribution of dopant ions are determined by the compatibility of the size and valance state of the ion compared to the host lattice ion it is replacing. As an example of the importance of symmetry in determining the properties of doped solids, the case of optically active ions in crystal hosts is described here. To introduce this topic, Sect. 4.1 provides a brief overview of the electronic properties of free ions. Following that, it is shown how group theory is used to determine the number and types of energy levels of these ions in different crystalline environments, and the selection rules for electronic transitions between their energy levels.

4.1 Energy Levels of Free Ions

The quantum mechanical system of interest is an ion consisting of a nucleus, electrons in filled inner shells, and electrons in unfilled outer shells. The latter are the "optically active" electrons that absorb or emit light while undergoing transitions between unfilled energy levels. The energy levels of the system are determined by the Coulomb interactions between the nucleus of the ion and each of the electrons, the Coulomb and exchange interactions among all of the electrons, and the spin–orbit interactions of the electrons. This problem is treated in detail in many quantum mechanics or atomic physics text books [1–3].

The free ion is in a physical environment of total spherical symmetry with both its energy and angular momentum being quantized. The interactions listed above determine the radial extent of the electron orbital, its shape, and its spatial

R.C. Powell, *Symmetry, Group Theory, and the Physical Properties of Crystals*,
Lecture Notes in Physics 824, DOI 10.1007/978-1-4419-7598-0_4,
© Springer Science+Business Media, LLC 2010

orientation. Each orbital represents an electronic state of the system described by an eigenfunction and an eigenvalue. In general the electronic states are degenerate.

The problem can be formulated in a center of mass coordinate system in which an optically active electron interacts with the nucleus as shielded by the inner shell electrons. The Hamiltonian for the system is

$$H = H_{KE} + H_{e-n} + H_{e-e} + H_{so}, \tag{4.1}$$

where the first term represents the kinetic energy of the electron, the second term is the electron's Coulomb interaction with the shielded nucleus, the third term contains the Coulomb and exchange interactions among all of the electrons in the unfilled outer shell, and the final term is the spin orbit interaction. The first two terms contribute an amount to the energy of the system that is common to all of the electronic energy levels. The interaction among the electrons in the unfilled shell plus the spin and orbital angular momentum contributions gives the relevant electronic energy levels of the ion. The Hamiltonian representing these contributions is

$$H_o = e^2 \sum_{i>j} r_{ij}^{-1} + \sum_i \xi(r_i) \vec{l}_i \cdot \vec{s}_i, \tag{4.2}$$

where e is the electronic charge, r_{ij} is separation of electrons i and j, $\xi(r_i)$ is the spin–orbit coupling parameter, $\vec{l}_i$ and $\vec{s}_i$ are the orbital and spin angular momentum vectors of the ith electron, and the sums run over all optically active electrons.

The single electron wave functions can be expressed as $\psi_i(\alpha n l m_l m_s) = |\alpha n l m_l m_s\rangle$ where n is the principal quantum number designating the energy of the state, l is the orbital angular momentum quantum, m_l is the orientational quantum number, m_s is the spin orientation quantum number, and α represents all other quantum numbers required to make a complete set for the system. Neglecting spin–orbit interaction and using only the first term in the Hamiltonian in (4.2), the results for the quantum mechanical operators acting on this wave function are

$$H_i|\alpha n l m_l m_s\rangle = E_n|\alpha n l m_l m_s\rangle, \tag{4.3}$$

$$\vec{l}_i^2|\alpha n l m_l m_s\rangle = l(l + 2soitis)\hbar^2|\alpha n l m_l m_s\rangle, \tag{4.4}$$

$$\vec{l}_{iz}|\alpha n l m_l m_s\rangle = m_l\hbar|\alpha n l m_l m_s\rangle, \tag{4.5}$$

$$\vec{s}_{iz}|\alpha n l m_l m_s\rangle = m_s\hbar|\alpha n l m_l m_s\rangle. \tag{4.6}$$

The total spin quantum number for an electron is always $s=1/2$ so it is not explicitly included in the expressions above. The spin angular momentum operator $\vec{s}_i$ is quantized so that $\vec{s}_i^2|\alpha n l m_l m_s\rangle = (3/4)\hbar^2|\alpha n l m_l m_s\rangle$. The spin orientation quantum number is $m_s = \pm 1/2$ representing spin up and spin down. The orbital angular momentum quantum l can have any integer value between 0 and $n-1$. The orientational quantum number m_l can have any integer value between $-l$ and $+l$. So every electronic energy state of the ion is $2n(2l+1)$ degenerate depending on the spin and orbital angular momentum states.

Angular momentum raising and lowering operators are also important

$$\vec{l_i}^{\pm}|\alpha n l m_l m_s\rangle = \hbar[(l \mp m_l)(l \pm m_l + 1)]^{1/2}|\alpha n l m_l \pm 1 m_s\rangle, \tag{4.7}$$

$$\vec{s_i}^{\pm}|\alpha n l m_l m_s\rangle = \hbar[(s \mp m_s)(s \pm m_s + 1)]^{1/2}|\alpha n l m_l m_s \pm 1\rangle, \tag{4.8}$$

where

$$\vec{l_i}^{\pm} = \vec{l}_{ix} \pm i \vec{l}_{iy}, \tag{4.9}$$

$$\vec{s_i}^{\pm} = \vec{s}_{ix} \pm i \vec{s}_{iy}. \tag{4.10}$$

These act to change either the orbital orientation state or the spin orientation state by ± 1.

In general, the strength of spin–orbit interaction is small compared to the Coulomb interaction described above. Thus it can be treated as a perturbation of the electronic energy levels that partially lifts the degeneracy through spin–orbit splitting. The second term in (4.2) can be rewritten in terms of the angular momentum operators as

$$H_{so} = \sum_i \xi_i \vec{l_i} \cdot \vec{s_i} = \sum_i \xi_i \left(\vec{j_i}^2 - \vec{l_i}^2 - \vec{s_i}^2 \right)/2, \tag{4.11}$$

where the total angular momentum is the vector sum of the orbital and spin angular momentum operators, $\vec{j} = \vec{l} + \vec{s}$, with quantum number j. The components of $\vec{j}$ can also be used to construct a raising and lowering operator for total angular momentum states. Using first-order perturbation theory, the new energy eigenvalues and eigenfunctions for the single electron ion considered above are

$$E_i = E_i^0 + \langle \psi_i^o | H_{so} | \psi_i^o \rangle \tag{4.12}$$

and

$$\psi_i = \psi_i^0 + \sum_{j \neq i} \frac{\langle \psi_j^o | H_{so} | \psi_i^o \rangle}{E_i^0 - E_j^0} \psi_j^o, \tag{4.13}$$

respectively. Using (4.11) for the Hamiltonian in (4.12) and the properties of the raising and lowering operators in (4.7) and (4.8) gives the additional energy of spin–orbit interaction:

$$E_{so} = \langle nljm_j|\xi(r)(j^2 - l^2 - s^2)/2|nl'j'm_j'\rangle$$

$$= \frac{\xi_{nl}}{2}[\,j(j+1) - l(l+1) - s(s+1)]\delta_{ss'}\delta_{ll'}\delta_{jj'}\delta_{m_jm_j'}. \qquad (4.14)$$

The spin–orbit coupling constant $\xi_{nl} = \langle nl|\xi(r)|nl\rangle$ depends on the radial extent of the electron orbital. Equation (4.14) shows that spin–orbit interaction splits the degenerate energy levels with total angular momentum quantum number j into a set of energy levels with j ranging from $|l-s|$ to $l+s$ in integer steps.

In spectroscopic notation, the orbital angular momentum quantum number of an electron is designated by a letter such that s represents $l=0$, p represents $l=1$, d represents $l=2$, f represents $l=3$, g represents $l=4$, etc. The electron configuration of optically active electrons on an ion is written as nl^m where n is the principal quantum number for the orbitals, m is the number of optically active electrons, and l is replaced by the letter designated described above. For example, nd^3 represents an ion configuration with three electrons in $l=2$ orbitals.

For ions with more than one optically active electron, the Coulomb, exchange, and spin–orbit interactions between all pairs of electrons must be taken into account. The electron wave functions are written as linear combinations of the products of single electron wave functions. These are constructed to be antisymmetric with respect to the interchange of two electrons in different orbitals to insure that the Pauli exclusion principle is satisfied. Since for our purposes we are interested in the symmetry properties of the system, we only need to consider the properties of angular momentum coupling. There are two possible situations to consider. The first is the case where the individual electron spin–orbit interaction is small compared to the Coulomb interaction between pairs of electrons. In this case the orbital angular momentum vectors of the individual electrons add vectorally to give the total angular momentum vector for the ion designated by $\vec{L}$,

$$\vec{L} = \sum_i \vec{l}_i \qquad (4.15)$$

and the spin angular momentum vectors of the individual electrons add vectorally to give the total spin angular momentum vector for the ion designated by

$$\vec{S} = \sum_i \vec{s}_i. \qquad (4.16)$$

Then spin–orbit interaction is accounted for by vectorally adding the total orbital angular moment and total spin angular momentum to obtain the total angular momentum for the ion designated $\vec{J}$;

$$\vec{J} = \vec{L} + \vec{S}. \qquad (4.17)$$

In the second case, spin–orbit interaction for an individual electron is stronger than the Coulomb interaction between pairs of electrons. In this case the individual electron total angular momenta are first calculated from the vector addition

$$\vec{j}_i = \vec{l}_i + \vec{s}_i \tag{4.18}$$

and then the total angular momentum for ion is found from the vector sum

$$\vec{J} = \sum_i \vec{j}_i. \tag{4.19}$$

Note that since the spin of an electron is ½, the total spin angular momentum S and the total angular momentum J can have half-integer values.

The strength of the spin–orbit interaction determines which of the two cases described above is appropriate. In either case the properties of angular momentum operators including the raising and lowering operators are the same for states with multielectron quantum numbers as they were for states with single electron quantum numbers. The quantum numbers designating the state of a multielectron ion are L, S, M_L, M_S, J, and M_J. The designation of an electronic state is given by $^{(2S+1)}L_J$ where the orbital angular momentum quantum number is replaced by the appropriate letter from the spectroscopic notation designated discussed above. The superscript $(2S+1)$ is the spin multiplicity. If its value is 1, 2, 3, etc., the state is referred to as a singlet, doublet, triplet, etc., respectively. A spectroscopic term is designated by the spin multiplicity and the letter for the orbital angular momentum. A multiplet of this term also includes the total angular momentum quantum number as subscript. For example, $^3P_2$ represents the $J=2$ multiplet of a triplet term ($S=1$) with orbital angular momentum $L=1$.

Since the electron–electron interaction Hamiltonian in (4.2) depends on r_{ij}^{-1}, it is useful to express the spatial part of the electron wave functions in terms of spherical harmonics and to use the usual expansion for r_{ij}^{-1} in terms of spherical harmonics:

$$\psi(j) = R_{nl}(j)Y_l^m(\theta, \varphi) \tag{4.20}$$

$$r_{ij}^{-1} = \sum_k \sum_{m=-k}^{k} \frac{4\pi}{2j+1} \frac{r_<^k}{r_>^{k+1}} Y_k^m(\theta_1, \varphi_1)Y_k^{m^*}(\theta_2, \varphi_2). \tag{4.21}$$

Here R_{nl} is the radial part of the wave function that depends on the principal quantum number and the orbital angular momentum quantum number. The angular part of the wave function is contained in the spherical harmonic function Y_l^m. The first few spherical harmonic functions are given in Table 4.1. It can be seen by inspection of the functions in the table that changing $\mathbf{r}$ to $-\mathbf{r}$ introduces a factor of $(-1)^l$. Thus $Y_l^m(\theta, \varphi)$ has even or odd parity depending on whether the angular momentum quantum number l is even or odd. The energy associated with electron–electron interactions is then found by evaluating the matrix element of the Hamiltonian expressed in terms of spherical harmonics and the electron wave functions expressed

Table 4.1 Spherical harmonic functions

$$Y_0^0 = \sqrt{\frac{1}{4\pi}}$$

$$Y_1^0 = \sqrt{\frac{3}{4\pi}}\frac{z}{r} = \sqrt{\frac{3}{4\pi}}\cos\theta$$

$$Y_1^{\pm 1} = \mp\sqrt{\frac{3}{8\pi}}\frac{x\pm iy}{r} = \mp\sqrt{\frac{3}{8\pi}}\sin\theta e^{\pm i\varphi}$$

$$Y_2^0 = \sqrt{\frac{5}{16\pi}}\frac{3z^2 - r^2}{r^2} = \sqrt{\frac{5}{16\pi}}(3\cos^2\theta - 1)$$

$$Y_2^{\pm 1} = \mp\sqrt{\frac{15}{8\pi}}\frac{z(x\pm iy)}{r^2} = \mp\sqrt{\frac{15}{8\pi}}\sin\theta\cos\theta e^{\pm i\varphi}$$

$$Y_2^{\pm 2} = \sqrt{\frac{15}{32\pi}}\frac{(x\pm iy)^2}{r^2} = \sqrt{\frac{15}{32\pi}}\sin^2\theta e^{\pm 2i\varphi}$$

$$Y_3^0 = \sqrt{\frac{7}{16\pi}}\frac{z(5z^2 - 3r^2)}{r^3} = \sqrt{\frac{7}{16\pi}}(5\cos^3\theta - 3\cos\theta)$$

$$Y_3^{\pm 1} = \mp\sqrt{\frac{21}{64\pi}}\frac{(x\pm iy)(5z^2 - r^2)}{r^3} = \mp\sqrt{\frac{21}{64\pi}}\sin\theta(5\cos^2\theta)e^{\pm i\varphi}$$

$$Y_3^{\pm 2} = \sqrt{\frac{105}{32\pi}}\frac{z(x\pm iy)^2}{r^3} = \sqrt{\frac{105}{32\pi}}\sin^2\theta(\cos\theta)e^{\pm 2i\varphi}$$

$$Y_3^{\pm 3} = \mp\sqrt{\frac{35}{64\pi}}\frac{z(x\pm iy)^3}{r^3} = \mp\sqrt{\frac{35}{64\pi}}\sin^3\theta e^{\pm 3i\varphi}$$

$$Y_4^0 = \sqrt{\frac{9}{256\pi}}\frac{35z^4 - 30z^2r^2 + 3r^4}{r^4} = \sqrt{\frac{9}{256\pi}}(35\cos^4\theta - 30\cos^2\theta + 3)$$

$$Y_4^{\pm 1} = \mp\sqrt{\frac{45}{64\pi}}\frac{(x\pm iy)(7z^3 - 3zr^2)}{r^4} = \mp\sqrt{\frac{45}{64\pi}}\sin\theta(7\cos^3\theta - \cos\theta)e^{\pm i\varphi}$$

$$Y_4^{\pm 2} = \sqrt{\frac{45}{128\pi}}\frac{(x\pm iy)^2(7z^2 - r^2)}{r^4} = \sqrt{\frac{45}{128\pi}}\sin^2\theta(7\cos^2\theta - 1)e^{\pm 2i\varphi}$$

$$Y_4^{\pm 3} = \mp\sqrt{\frac{315}{64\pi}}\frac{z(x\pm iy)^3}{r^4} = \mp\sqrt{\frac{315}{64\pi}}\sin^3\theta\cos\theta e^{\pm 3i\varphi}$$

$$Y_4^{\pm 4} = \sqrt{\frac{315}{512\pi}}\frac{(x\pm iy)^4}{r^4} = \sqrt{\frac{315}{512\pi}}\sin^4\theta e^{\pm 4i\varphi}$$

in terms of spherical harmonics. Note that for the matrix element to be nonzero the product of the spherical harmonics involved must be an even function.

For optical absorption or emission transitions between energy levels, the radiation field can be expressed in terms of a multipole expansion with the leading term being the electric dipole term

$$H_{ed} = e\vec{r}. \tag{4.22}$$

As discussed in Sect. 2.4, a quantum mechanical transition between an initial and final state of a system is proportional to the matrix element describing the process. For an optical transition induced by the electric dipole term of the radiation field (neglecting polarization) the matrix element can be expressed as

$$M_{ed} = \left\langle \psi_f^e \left| e\vec{r} \right| \psi_i^e \right\rangle. \tag{4.23}$$

If this matrix element is identically zero the transition is said to be forbidden while if it is nonzero the transition is allowed. Since the dipole moment operator is an odd function that can be expressed in terms of spherical harmonics with $l=1$, the initial and final states of the system must have opposite parity for the transition to be allowed. Thus all optical transitions between single electron states with the same angular momentum quantum number are forbidden.

Since free ions are in an environment of totally spherical symmetry, group theory is not a helpful tool. However, if an external perturbation such as an electric or magnetic field is applied with specific directional properties, group theory is useful in determining change in energy levels and the selection rules for optical transitions. This is what occurs when an ion is doped into a crystal lattice as described in Sect. 4.2.

4.2 Crystal Field Symmetry

When an ion with optically active electrons is doped into a crystalline host lattice, the Hamiltonian describing its energy levels is given by (4.1) with an additional term H_{cf} added to take into account the effect of the crystal field. The dopant ion is no longer in an environment of spherical symmetry but rather is surrounded with a set of host ions located at specific nearest-neighbor lattice positions. These ions are referred to as ligands. Each ligand has an electric field associated with it, and thus the dopant ion finds itself in an environment of an electric field with a specific geometric shape. H_{cf} describes the interaction of the optically active electrons with this electric field. There are three approaches to treating this problem depending on the relative strengths of the interactions described by the Hamiltonians in (4.1) compared to the strength of the crystal field interaction [4–6]. These are described below.

If the crystal field interaction is small compared to both the electronic Coulomb interaction and the spin–orbit interaction, the situation is referred to as a *weak crystal field case*. This case is treated using the free-ion multiplets described by total angular momentum quantum numbers J and M_J for the eigenfunctions of the unperturbed system and treating the crystal field as a perturbation on the system. This perturbation causes a *Stark splitting* of the free-ion multiplets and determines the selection rules for optical transitions between any two split energy levels. Rare earth ions are an example of a weak crystal field case since their 4f optically active electrons are shielded from the crystal field by other electrons in different orbitals with greater radii. An example of this is discussed in Sect. 4.5.

The opposite situation is one in which the magnitude of the crystal field interaction is greater than either the electron Coulomb interaction or the spin–orbit interaction and thus is called a *strong crystal field case*. In this case the eigenfunctions of the unperturbed system are taken to be the single-electron wave functions designated by quantum numbers l, m_l, s, and m_s. The crystal field acts as a perturbation to split these energy levels into a set of crystal field states. Then multielectron terms are formed by taking into account the effect of electron Coulomb interactions, and finally crystal field multiplets are determined from spin–orbit interaction. Good examples of strong crystal field case can be found in second and third row transition metal ions. This is discussed further in Sect. 4.4.

The *medium crystal field case* occurs when the strength of the crystal field is greater than the magnitude of spin–orbit interaction but less than the strength of the electron Coulomb interaction. In this case the eigenfunctions of the unperturbed system are the free ion terms designated by the quantum numbers L, S, M_L, and M_S. The effect of the crystal field is to split these energy levels into a set of crystal field terms. Then spin–orbit interaction is applied to form crystal field multiplets. The first row transition metal ions with unshielded 3d optically active electrons are good examples of a medium crystal field case. However, it is more difficult to work with this case so it is common to treat the crystal field effects on 3d ions using either the strong field or weak field schemes. The results come out the same.

The crystal field Hamiltonian, the wave functions of free ions, and the electric dipole operator can all be expressed in terms of spherical harmonics. Group theory can be used to determine how these functions transform under specific symmetry operations. The angular parts of these functions are given by

$$Y_l^m = NP_l^m(\cos\theta)e^{im\varphi}, \tag{4.24}$$

where the $P_l^m(\cos\theta)$ are the Legendre polynomials and N is a normalization factor, as shown in Table 4.1. The splitting of the energy levels in the crystal field is determined from perturbation theory using

$$E_l \propto \langle\psi_l|H_{cf}|\psi_l\rangle \tag{4.25}$$

and then the electronic transition between levels is determined from (4.23) using the crystal field energy levels and the irreducible representation for the dipole moment operator.

The first step in treating the effects of a crystal field on an optically active ion is to treat the ligands as point charges and determine the electric field at the site of the ion. To do this the number and location of the ligands with respect to the central ion must be designated. As an example, consider the octahedral coordination of six ligands each with a charge Ze located as shown in Fig. 4.1. This situation was treated in the example of O_h point group symmetry in Sect. 2.3.1. The difference here is that the central ion is not the same as the other six ions. The crystal field Hamiltonian is found from the electrostatic field at the positions of the optically active electrons

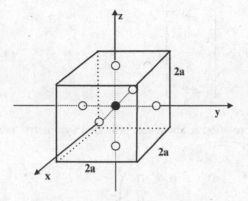

ION POSITIONS	
x,y,z,	r,θ,φ
Optically active ion:	
0,0,0	0,0,0
Ligands:	
a,0,0	a,π/2,0
0,a,0	a,π/2,π/2
−a,0,0	a,π/2/π
0,-a,0	a,π/2,3π/2
0,0,a	a,0,0
0,0,−a	a,π,0

Fig. 4.1 Central ion with octahedral coordination of ligands

$$H_{cf} = \sum_j eV_{cf}\left(r_j,\theta_j,\varphi_j\right) = \sum_j e\sum_{i=1}^{6}\frac{eZ}{|r_j - r_i|}$$

$$= \sum_j e\sum_{i=1}^{6}\sum_{l=0}^{\infty}\sum_{m=-l}^{l}\frac{4\pi eZ}{2l+1}\frac{r^l}{a^{l+1}}Y_l^{m*}(\theta_i,\varphi_i)Y_l^m\left(\theta_j,\varphi_j\right).$$

(4.26)

Here the standard multipole expansion for $|r_j - r_i|^{-1}$ with $|r_j|<|r_i|=a$ has been used so the crystal field can be expressed in terms of spherical harmonic functions. The sum over six ligands is consistent with the example of octahedral coordination.

For d electrons with $l=2$, the orthogonality of spherical harmonics implies that the product of the two eigenfunctions in (4.25) will result in spherical harmonics with $l \leq 4$. Thus the crystal field matrix element will be zero for any spherical harmonic terms in H_{cf} with l greater than 4. Similarly, for f electrons with $l=3$, the terms in H_{cf} that give nonzero terms in the crystal field matrix element must have $l\leq6$. In addition, H_{cf} must contain only even parity terms for the matrix element to be nonzero.

A specific example of a d electron in an octahedral crystal field is described in Sect. 4.3.

4.3 Energy Levels of Ions in Crystals

Consider an ion in a crystal field of a specific symmetry. If the system is quantized along a major axis of symmetry, a rotation of α about this axis changes the exponential factor of a spherical harmonic function from $e^{im\phi}$ into $e^{im\phi + \alpha}$. The ion has an orbital angular momentum quantum represented by number l and its wave function is given by (4.20). Its orientation quantum number ranges from $m_l=+l$ to $m_l=-l$ in integral steps. A linear combination of these $2l+1$ wave functions forms the basis function for a representation of the symmetry group of the system [7]. This can be expressed as

$$\Psi_l \propto \begin{bmatrix} e^{il\varphi} \\ e^{i(l-1)\varphi} \\ \vdots \\ e^{-il\varphi} \end{bmatrix}. \tag{4.27}$$

The symmetry operation of a rotation α about the principal symmetry axis is represented by the matrix

$$R = \begin{pmatrix} e^{il\alpha} & 0 & 0 & 0 \\ 0 & e^{i(l-1)\alpha} & \vdots & 0 \\ \vdots & \vdots & \ddots & \vdots \\ 0 & 0 & 0 & e^{-il\alpha} \end{pmatrix}. \tag{4.28}$$

The multiplication of (4.28) and (4.27) transforms the wave function as required.

The character of the α rotation operator is found by taking the trace of the rotation matrix in (4.28):

$$\chi(\alpha) = \mathrm{Tr}(R) = e^{il\alpha} + e^{i(l-1)\alpha} + \cdots + e^{-il\alpha} = \frac{\sin\left[(l+\frac{1}{2})\alpha\right]}{\sin\left(\frac{1}{2}\alpha\right)}. \tag{4.29}$$

For example, the characters for some of the common rotation operations are

$$\alpha = \pi, \qquad \chi(C_2) = (-1)^l$$

$$\alpha = \pi/2, \qquad \chi(C_4) = \begin{cases} 1, & l = 0, 1, 4, 5, \cdots \\ -1, & l = 2, 3, 6, 7, \cdots \end{cases}$$

$$\alpha = 2\pi/3, \qquad \chi(C_3) = \begin{cases} 0, & l = 1, 4, \cdots \\ 1, & l = 0, 3, 6, \cdots \\ -1, & l = 2, 5, \cdots \end{cases}$$

Thus, for a d electron with $l=2$, the characters for rotation operations are $\chi(C_2)=1$, $\chi(C_4)=-1$, and $\chi(C_3)=-1$. For an f electron with $l=3$, the characters are $\chi(C_2)=-1$, $\chi(C_4)=-1$, and $\chi(C_3)=1$.

For the case where spin–orbit interaction is stronger than the crystal field, the free-ion wave functions are designated by the total angular momentum quantum number j instead of the orbital angular momentum quantum number l. In this case the expression for the character of an operation of rotation through an angel α given in (4.29) becomes

$$\chi(\alpha) = \frac{\sin\left[\left(j + \frac{1}{2}\right)\alpha\right]}{\sin\left(\frac{1}{2}\alpha\right)}. \tag{4.30}$$

When l or j has integer values, $\chi(\alpha+2\pi) = \chi(\alpha)$ so a rotation by 2π is the identity operation. However, when an atom has an odd number of electrons, j can also have half-integer values. When this occurs, (4.30) shows that $\chi(\alpha+2\pi)=-\chi(\alpha)$ is not the identity operator but rather a new symmetry operation designated by R. In this situation the character of the identity operation is given by

$$\chi(E) = 2j + 1 \tag{4.31}$$

and

$$\chi(R) = -(2j + 1). \tag{4.32}$$

This case is called a *double-valued representation* as discussed in Sect. 2.2. The number of symmetry operations in the group is now twice the original number and includes the products of R with all of the other members of the group.

To determine the transformation matrix for nonrotational symmetry elements such as reflection and inversion, it is necessary to consider how the x, y, and z components of the spherical harmonics given in Table 4.1 transform. Then the traces of these matrices give the characters of the operations instead of (4.29) or (4.30).

The optically active ions of most interest [4] have orbital angular momentum quantum numbers of $l=2$ or 3. As an example of the use of group theory, consider a strong crystal field case for a d-electron in an octahedral crystal of symmetry class O. The normal character table for this point group is given in Table 2.31. In order to treat spin–orbit interaction, this is expanded to the character table for the full double group in the top portion of Table 4.2 while some special representations and their reductions are shown in the bottom portion of the table. Using this table and (4.29) and (4.30) gives the characters for the Γ_2 representation as shown in the lower half of Table 4.2. Here Γ_l is the crystal field symmetry state for ions with orbital angular momentum l (in this case 2). The Γ_2 reducible crystal field representation has been reduced in terms of the irreducible representations of the O group using (2.10). This shows that the fivefold orbitally degenerate free ion energy level of a d-electron in a crystal field of O symmetry splits into two crystal field energy levels, a doubly degenerate E level and a triply degenerate T level as depicted in Fig. 4.2. The orbital angular momentum l is no longer a good quantum number. It has been replaced by the irreducible representation designations in the crystal field. Physically this means that the electron is not free to move about the nucleus of the ion as dictated by internal Coulomb interactions, but instead its motion is constrained by the crystal field established by the ligands. Note that the spin multiplicity remains unchanged since the new spatial symmetry does not change the spin state of the ion. Thus the total degeneracy of these states is twice the orbital degeneracy as listed in Fig. 4.2. This analysis is the same if the unperturbed ion is represented by the total angular

Table 4.2 Representations in a crystal field with O symmetry

O	E	$8C_3$	$3C_2$	$6C_4$	$6C'_2$	R	$8RC_3$	$3RC_2$	$6RC_4$	$6RC'_2$	
A_1	1	1	1	1	1	1	1	1	1	1	
A_2	1	1	-1	-1	1	1	1	-1	-1	1	
E	2	-1	0	0	2	2	-1	0	0	2	
T_1	3	0	-1	1	-1	3	1	0	-1	-1	(x,y,z)
T_2	3	0	1	-1	-1	3	-1	0	-1	1	
$D_{1/2}$	2	1	0	$\sqrt{2}$	0	-2	-1	0	$-\sqrt{2}$	0	
$_2S$	2	1	0	$-\sqrt{2}$	0	-2	-1	0	$\sqrt{2}$	0	
$D_{3/2}$	4	-1	0	0	0	-4	1	0	0	0	
Γ_2	5	-1	1	-1	1	5	-2	0	-1	3	$= E + T_2$
Γ_4	9	0	1	1	1	9	0	1	-1	1	$=A_1+E+T_1+T_2$
$\Gamma_{5/2}$	6	0	0	$-\sqrt{2}$	0	-6	0	0	$\sqrt{2}$	0	$=_2S+D_{3/2}$
$\Gamma_{3/2}$	4	-1	0	0	0	-4	1	0	0	0	$=D_{3/2}$
$\Gamma_{1/2}$	2	1	0	$\sqrt{2}$	0	-2	-1	0	$-\sqrt{2}$	0	$=D_{1/2}$
$D_{1/2}\times T_2$	6	0	0	$-\sqrt{2}$	0	-6	1	0	$\sqrt{2}$	0	$=_2S+D_{3/2}$
$D_{1/2}\times E$	4	-1	0	0	0	-4	1	0	0	0	$= D_{3/2}$

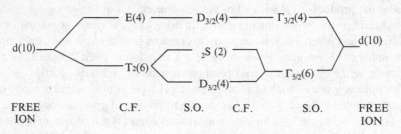

Fig. 4.2 Splitting of the energy level of a d-electron in a strong and weak crystal field of O symmetry (The numbers in parentheses indicate the degeneracy of the level.)

momentum quantum number $j=2$ or a Russell–Saunders term with $L=2$ as discussed in Sect. 4.1.

The next order of perturbation is spin–orbit interaction. The spin state of a single electron with $s=1/2$ transforms as the representation $\Gamma_{1/2}$ in the O symmetry group. This has the characters derived by (4.30) shown in Table 4.2. From inspection (or using (2.10)), the $\Gamma_{1/2}$ reducible representation transforms as the $D_{1/2}$ irreducible representation. Taking the product of this representation with the orbital state T_2 and E representations gives the representations of the spin–orbit coupled states in the octahedral crystal field. As shown in Table 4.2 and Fig. 4.2, the use of (2.10) shows that the E state becomes a $D_{3/2}$ state while the T_2 state splits into a $_2S$ state and a $D_{3/2}$ state. Thus the tenfold degenerate free ion d energy level becomes two fourfold degenerate $D_{3/2}$ levels and one twofold degenerate $_2S$ level in a crystal field with O symmetry and spin–orbit coupling.

For strong spin–orbit coupling, the free ion with a d electron is split into two levels with total angular momentum quantum numbers $j=5/2$ or $3/2$. Equation (4.30) can be used to determine the characters of the irreducible representation

$\Gamma_{5/2}$ and $\Gamma_{3/2}$ in the O symmetry group as shown in Table 4.2. These can be reduced in term of the $_2S$ and $D_{3/2}$ irreducible representations. The $\Gamma_{5/2}$ state undergoes a crystal field splitting into two levels while the $\Gamma_{3/2}$ state does not split. Note that the crystal field states are the same whether spin–orbit interaction is accounted for before or after crystal field splitting.

This example shows the ability of group theory to determine the manner in which the energy levels of free ions are split by crystal field and spin–orbit perturbations. This is done using very simple manipulations of character tables and the properties of irreducible representations without the need for very complicated quantum mechanical perturbation theory calculations. However, it should be noted that simple symmetry arguments cannot provide information on the quantitative magnitude of the energy level splittings.

Group theory can also be used to determine whether or not electronic transitions between energy levels are allowed or forbidden. It was shown by (4.23) that electric dipole transitions between free ions states with the same l quantum number are forbidden. However, since l is not a good quantum number in the crystal, the electric dipole matrix element describing the transition must be considered between states designated by crystal field representations. Transitions that are allowed in the crystal are called *forced electric dipole transitions*. Allowed transitions of this type appear as strong lines in the absorption and emission spectra of the material while forbidden transitions produce much weaker spectral features.

As discussed in Sect. 2.4, when the initial and final states of the transition are designated by the irreducible representations of the crystal field symmetry group Γ_i and Γ_f and the operator of the radiation field causing the transition is designated by the irreducible representation Γ_{ed}, (4.23) for the transition matrix element can be rewritten in terms of the products of irreducible representations. To determine the selection rules for the transitions, (4.23) is now expressed as

$$\Gamma_i \times \Gamma_{ed} \times \Gamma_f \supset A_{1g} \quad \text{or} \quad \Gamma_i \times \Gamma_{ed} \supset \Gamma_f. \tag{4.33}$$

The first expression states that the reduction of the triple product irreducible representations must contain the totally symmetric representation of the crystal field group for the transition to be allowed. As discussed in Sect. 2.4, this is because the matrix element involves an integral of the product of these three functions over all space and this integral will be identically zero unless the integrand is a totally symmetric function. The second expression in (4.33) follows from the fact that the totally symmetric representation only appears in the direct product of an irreducible representation with itself. Thus the reduction of the direct product of any two of the irreducible representations involved in the transition matrix element must contain the third irreducible representation for the transition to be allowed.

In the example above with O symmetry, the electric dipole momentum operator transforms as T_1. Thus for the example of O symmetry, it is necessary to find the direct product representations of T_1 with each of the irreducible representations of the group and reduce them in terms of the irreducible representations. Using the characters in Table 4.2 and (2.10) gives

$$T_1 \times A_1 = T_1$$
$$T_1 \times A_2 = T_2$$
$$T_1 \times E = T_1 + T_2$$
$$T_1 \times T_1 = T_1 + T_2 + E + A_1$$
$$T_1 \times T_2 = T_1 + T_2 + E + A_2$$
$$T_1 \times D_{1/2} = D_{1/2} + D_{3/2}$$
$$T_1 \times {_2S} = {_2S} + D_{3/2}$$
$$T_1 \times D_{3/2} = 2D_{3/2} + {_2S} + D_{1/2}.$$

From this analysis it can be seen that transitions between the crystal field levels T_2 and E in Fig. 4.2 are allowed. Also transitions between all of the spin–orbit split levels in Fig. 4.2 are allowed.

Group theory can also be used to determine the exact form of the crystal field Hamiltonian [4, 6]. As discussed in Sect. 4.2, the expression for H_{cf} in (4.26) has a limited number of terms in the sum over l depending on the orbital angular momentum of the optically active electron. For d electrons, only the $l=2,4$ terms are nonzero. Using the technique described above, Table 4.2 shows how the Γ_2 and Γ_4 reducible representations transform in O symmetry and their reduction in terms of the irreducible representations of the group. Since H_{cf} is part of the total Hamiltonian of the system, it must transform as the A_1 irreducible representation in order to remain invariant under all symmetry operations of the group. Table 4.2 shows that there are no linear combinations of the Y_2^m functions that transform as A_1 but there is a linear combination of Y_4^m functions that transform as A_1. Thus the crystal field expansion in (4.26) contains only Y_4^m terms.

To determine the specific linear combination of Y_4^m terms in the expansion of H_{cf}^O, symmetry operations of the group must be applied to the set of nine spherical harmonic functions with $m = -4,\ldots +4$. In Cartesian coordinates, a rotation of $\pi/2$ about the z-axis will take (x,y,z) into $(y,-x,z)$. Applying this to the spherical harmonic functions in Table 4.1 gives

$$C_4 \begin{pmatrix} Y_4^4 \\ Y_4^3 \\ Y_4^2 \\ Y_4^1 \\ Y_4^0 \\ Y_4^{-1} \\ Y_4^{-2} \\ Y_4^{-3} \\ Y_4^{-4} \end{pmatrix} = \begin{pmatrix} Y_4^4 \\ iY_4^3 \\ -Y_4^2 \\ -iY_4^1 \\ Y_4^0 \\ iY_4^{-1} \\ -Y_4^{-2} \\ -iY_4^{-3} \\ Y_4^{-4} \end{pmatrix}.$$

Since H_{cf}^O must remain unchanged under the C_4 operation, the only nonzero expansion coefficients in (4.26) are those for Y_4^0 and $Y_4^{\pm 4}$.

Applying the C'_2 changes (x,y,z) into $(y,x,-z)$. Y_4^0 is invariant under this operation while Y_4^4 and Y_4^{-4} transform into each other. Thus,

$$H_{cf}^O \propto Y_4^0 + d(Y_4^4 + Y_4^{-4}).$$

The coefficient d can be evaluated by applying the C_3 operation of the O group. This operation takes (x,y,z) into (y,z,x) and thus mixes all three spherical harmonics. Since

$$C_3 H_{cf}^O \equiv H_{cf}^O,$$

$$(35x^4 - 30x^2r^2 + 3r^4)/8 + d\sqrt{\frac{35}{128}}\left[(y+iz)^4 + (y-iz)^4\right]$$

$$= (35z^4 - 30z^2r^2 + 3r^4)/8 + d\sqrt{\frac{35}{128}}\left[(x+iy)^4 + (x-iy)^4\right].$$

Since these must be identical, the coefficients of z^4 on both sides of the equation must be equal:

$$\frac{3}{8} + 2d\sqrt{\frac{35}{128}} = 1.$$

Thus, $d = \sqrt{5/14}$ so the form of the octahedral crystal field becomes

$$H_{cf}^O \propto Y_4^0 + \sqrt{\frac{5}{14}}(Y_4^4 + Y_4^{-4}). = x^4 + y^4 + z^4 - \frac{3}{5}r^4. \qquad (4.34)$$

Now that the form of the crystal field Hamiltonian is known for octahedral symmetry, (4.25) can be used to determine the magnitude of the free ion energy level splittings due to the presence of H_{cf}^O. For the example of ions with optically active 3d electrons, the wave functions are

$$\psi_{3,2,m_l} = R_{32}(r)Y_2^{m_l}(\theta,\varphi). \qquad (4.35)$$

So,

$$\psi_{3,2,0} = R_{3d}\sqrt{\frac{1}{2\pi}}\sqrt{\frac{5}{8}}(3\cos^2\theta - 1)$$

$$\psi_{3,2,\pm 1} = \mp R_{3d}\sqrt{\frac{1}{2\pi}}\sqrt{\frac{15}{4}}(\cos\theta\sin\theta e^{\pm i\varphi})$$

$$\psi_{3,2,\pm 2} = R_{3d}\sqrt{\frac{1}{2\pi}}\sqrt{\frac{15}{16}}(\sin^2\theta e^{\pm i2\varphi}).$$

and

$$H_{cf}^O = C(R)\sqrt{\frac{1}{2\pi}}$$

$$\times \left[\sqrt{\frac{9}{128}}(35\cos^4\theta - 30\cos^2\theta + 3) + \sqrt{\frac{5}{14}}\sqrt{\frac{315}{256}}\sin^4\theta(e^{i4\varphi} - e^{-i4\varphi}) \right].$$

Here $C(R)$ is a factor containing all of the expansion coefficients and radial factors.

The crystal field matrix elements involve integrals over θ and φ. Considering the φ integral first,

$$M = \langle\psi_i|H_{cf}^O|\psi_j\rangle \propto \int_0^{2\pi} (e^{im_i\varphi})^* e^{im_O\varphi} e^{im_j\varphi} d\varphi = \int_0^{2\pi} e^{i(m_O+m_j-m_i)\varphi} d\varphi.$$

This integral will equal zero unless $m_O+m_j-m_i=0$. Since m_O has the values 0, ± 4, $\Delta m_{ij}=0$, ± 4. The values of m_i and m_j for d-electrons are 0, ± 1, ± 2. This shows that all diagonal matrix elements ($\Delta m_{ij}=0$) are nonzero and matrix elements with $m_i=\pm 2$, $m_j=\pm 2$ are nonzero. These are given by

$$\langle\psi_{3,2,0}|H_{cf}^O|\psi_{3,2,0}\rangle = K\frac{5}{8}\sqrt{\frac{9}{128}}\int_0^\pi (3\cos^2\theta - 1)(35\cos^4\theta - 30\cos^2\theta + 3)(3\cos^2\theta - 1)\sin\theta\,d\theta,$$

$$\langle\psi_{3,2,\pm1}|H_{cf}^O|\psi_{3,2,\pm1}\rangle = K\frac{15}{4}\sqrt{\frac{9}{128}}\int_0^\pi (\cos\theta\sin\theta)(35\cos^4\theta - 30\cos^2\theta + 3)(\cos\theta\sin\theta)\sin\theta\,d\theta,$$

$$\langle\psi_{3,2,\pm2}|H_{cf}^O|\psi_{3,2,\pm2}\rangle = K\frac{15}{16}\sqrt{\frac{9}{128}}\int_0^\pi (\sin^2\theta)(35\cos^4\theta - 30\cos^2\theta + 3)(\sin^2\theta)\sin\theta\,d\theta,$$

$$\langle\psi_{3,2,\pm2}|H_{cf}^O|\psi_{3,2,\mp2}\rangle = K\frac{15}{16}\sqrt{\frac{5}{14}}\sqrt{\frac{315}{256}}\int_0^\pi (\sin^2\theta)(\sin^4\theta)(\sin^2\theta)\sin\theta\,d\theta.$$

Here all of the radial functions, expansion parameters, and factors of $\sqrt{2/\pi}$ have been into the parameter K. These integrals can be evaluated using the expression

$$\int_0^\pi (\sin^{2n+1}\theta)(\cos^m\theta)d\theta = \int_0^\pi (\sin^m\theta)(\cos^{2n+1}\theta)d\theta$$

$$= \frac{2^{n+1}n!}{(m+1)(m+3)\cdots(m+2n+1)}.$$

It is conventional to express that the results are in terms of the parameters D and q which are constants multiplied by the factor K. This gives

$$\langle \psi_{3,2,0} | H_{cf}^O | \psi_{3,2,0} \rangle = 6Dq$$
$$\langle \psi_{3,2,\pm1} | H_{cf}^O | \psi_{3,2,\pm1} \rangle = -4Dq$$
$$\langle \psi_{3,2,\pm2} | H_{cf}^O | \psi_{3,2,\pm2} \rangle = Dq$$
$$\langle \psi_{3,2,\pm2} | H_{cf}^O | \psi_{3,2,\mp2} \rangle = 5Dq.$$

The secular determinant with these elements is

$$
\begin{vmatrix}
Dq - E & 0 & 0 & 0 & 5Dq \\
0 & -4Dq - E & 0 & 0 & 0 \\
0 & 0 & 6Dq - E & 0 & 0 \\
0 & 0 & 0 & -4Dq - E & 0 \\
5Dq & 0 & 0 & 0 & Dq - E
\end{vmatrix}.
$$

This can be box diagonalized and expanded to give the secular equation

$$(-4Dq - E)^2 (6Dq - E) \lfloor (Dq - E)^2 - (5Dq)^2 \rfloor = 0.$$

The solution to this shows that there is a triply degenerate level with $E = -4Dq$ and a doubly degenerate level with $E = 6Dq$. Thus the crystal field splitting of the T_2 and E levels in Fig. 4.2 is $10Dq$.

The explicit expressions for D and q are found from evaluating the matrix elements using all of the radial functions, expansion parameters, and numerical factors. This gives

$$D = \frac{35e}{4a^5} \tag{4.36}$$

and

$$q = \frac{2Ze}{105} \langle R_{3,d}(r) | r^4 | R_{3,d}(r) \rangle. \tag{4.37}$$

This shows that the magnitude of the crystal field splitting $10Dq$ varies as the inverse fifth power of the ion–ligand distance.

4.4 Example: d-Electrons

Materials with transition metal ion dopants having optically active d-electrons are important in applications such as lasers and phosphors and in giving color to some gem stones [4, 5] One important example is ruby. The host material is aluminum oxide, Al_2O_3. This is a clear crystal known as sapphire. When less than 1% of the

Al^{3+} host ions are replaced by Cr^{3+} ions the color of the crystal turns to red and it is called ruby. Trivalent chromium has three optically active electrons in the unfilled 3d shell outside of a filled inner shell. Since these are unshielded, they are strongly affected by a crystal field.

In the sapphire host crystal, the Al^{3+} ions are surrounded by six oxygen ions producing a crystal field of almost O_h symmetry. This octahedron is slightly twisted producing a distortion that results in the actual symmetry being C_3. However, the strength of the octahedral part of the crystal field is so much stronger than the slight trigonal distortion that the major properties of the optical spectra can be explained by assuming O_h symmetry. Detailed fine structure can then be explained by treating the reduction in symmetry from O_h to C_3.

When a Cr^{3+} ion is substituted for an aluminum ion, the energy levels and electronic transitions of its optically active electrons can be described by treating three d electrons in an octahedral crystal field site. As described above, a free ion d electron energy level splits into a triply degenerate and a doubly degenerate energy in an octahedral crystal field. These crystal field energy levels are designated by the irreducible representations of the O_h symmetry group as discussed previously. It is common convention to use capital letters for the irreducible representations when dealing with multielectron energy terms as discussed above. However, for single electron states the convention is to use small letters for the irreducible representations. Thus, the multielectron T_2 and E levels shown in Fig. 4.2 are designated t and e when dealing with a single d-electron. In a addition, for groups with an inversion symmetry element, it is necessary to include the subscript g (gerade) or u (ungerade) to designate an even parity or odd parity representation. For a d-electron with angular momentum quantum number $l=2$, the wave functions have even parity so the single electron orbitals of interest are t_{2g} and e_g.

Figure 4.2 shows that the triply orbitally degenerate t_{2g} level will be of lower energy than the doubly orbitally degenerate e_g level. In the strong field model, all three electrons will occupy the lowest energy t_{2g} orbitals as shown in Fig. 4.3. The arrows indicate the direction of the electron spin. Hund's rules state that the lowest energy state will be the one with the highest spin multiplicity [4] Thus the ground state will have all three spins of the electrons aligned producing quartet terms. The configuration of single electrons is designated $t_{2g}^3 e_g^0$ where the superscripts indicate how many electrons are in each type of orbital.

There are two types of electronic transitions in this model as shown in Fig. 4.3. The first type is a spin-flip transition in which all the electrons retain the $t_{2g}^3 e_g^0$ configuration but now have a spin multiplicity of 2, giving doublet terms. Note that the energy related to flipping the spin of an electron is independent of the magnitude of the crystal field. These transitions appear as sharp lines in the optical absorption and emission spectra.

The second type of transition is one that changes the crystal field configuration from $t_{2g}^3 e_g^0$ to $t_{2g}^2 e_g^1$. As shown in Fig. 4.3, the energy of this type of transition depends on the magnitude of the crystal field. Since $10Dq$ varies as a^{-5}, this changes with the thermal vibrations of the crystal lattice that modulate the central ion–ligand separation. This thermal modulation of the crystal field imparts an

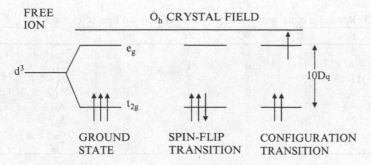

Fig. 4.3 Optical transitions of three d electrons in an O_h crystal field

energy uncertainty to this type of transition. Therefore configuration-changing transitions appear as broad bands in the optical absorption and emission spectra. The excited state for this case is still a quartet term since the total spin has not been changed.

In order to obtain the complete set of crystal field terms, two other single electron configurations must be considered along with the two given above. These are $t_{2g}e_g^2$ and $t_{2g}^0e_g^3$. The multielectron terms are found from group theory by finding the direct products representations of single electron configurations and reducing these in terms of the irreducible representations of the O_h symmetry group. As an example, consider the ground state configuration $t_{2g}^3e_g^0$. This requires taking the triple direct product $t_{2g} \times t_{2g} \times t_{2g}$. Using the character table for O_h given in Table 2.32, the products of the characters of the t_{2g} irreducible representation with itself give the characters of the $t_{2g} \times t_{2g}$ representation. This reducible representation can be reduced in terms of the irreducible representations $A_{1g}+E_g+T_{1g}+T_{2g}$ by using (2.10). The next step in evaluating the triple direct product is to find the direct product of t_{2g} with each of these irreducible representations and determine the reductions of their product representations. Again using Table 2.32 and (2.10) gives $t_{2g} \times A_{1g}=T_{2g}$, $t_{2g} \times E_g=T_{1g}+T_{2g}$, $t_{2g} \times T_{1g}=A_{2g}+E_g+T_{1g}+T_{2g}$, and $t_{2g} \times T_{2g}=A_{1g}+E_g+T_{1g}+T_{2g}$. Similar procedures for the other three single electron configurations result in twenty multielectron doublet and quartet terms for the Cr^{3+} ion in an O_h crystal field.

In the weak field model, the rules for coupling the angular momenta of three electrons each with quantum numbers ($n=3$, $l=2$, $m_l=\pm 2, \pm 1, 0$, $s=1/2$, $m_s=\pm 1/2$) can be used to determine the multielectron terms of the free ion. Forming all allowed combinations of quantum numbers shows that the free ion terms for the Cr^{3+} ion are two quartets and six doublets: $^4F, ^4P, ^2H, ^2G, ^2F, ^2D, ^2D, ^2P$.

When these free ion terms are put into a crystal field environment with O_h symmetry, they split into states designated by the irreducible representations of the group. This can be demonstrated more simply by using the O subgroup of O_h which gives the same results. A summary of the results is presented in Table 4.3.

Table 4.3 Reduction of the terms of d^3 free ions in a crystal field of O symmetry

O	E	$3C_2$	$8C_3$	$6C_4$	$6C'_2$	
^4F	7	-1	1	-1	-1	$^4A_{2g}+{}^4T_{1g}+{}^4T_{2g}$
^4P	3	-1	0	1	-1	$^4T_{1g}$
^2H	11	-1	-1	1	-1	$^2E_g+2{}^2T_{1g}+{}^2T_{2g}$
^2G	9	1	0	1	1	$^2A_{1g}+{}^2E_g+{}^2T_{1g}+{}^2T_{2g}$
^2F	7	-1	1	-1	-1	$^2A_{2g}+{}^2T_{1g}+{}^2T_{2g}$
^2D	5	1	-1	-1	1	$^2E_g+{}^2T_{2g}$
^2D	5	1	-1	-1	1	$^2E_g+{}^2T_{2g}$
^2P	3	-1	0	1	-1	$^2T_{1g}$

As an example of the results shown in Table 4.3, consider the ^2G term with $L=4$. The characters are found from (4.29) to be

$$\chi(E) = 2 \times 4 + 1 = 9 \qquad \chi(C_3) = \frac{\sin 3\pi}{\sin \frac{\pi}{3}} = 0$$

$$\chi(C_2) = \frac{\sin \frac{9\pi}{2}}{\sin \frac{\pi}{2}} = 1 \qquad \chi(C_4) = \frac{\sin \frac{9\pi}{4}}{\sin \frac{\pi}{4}} = 1.$$

Then (2.10) can be used to determine how many times each irreducible representation of the O group appears in the reduction of the ^2G reducible representation;

$$n(A_1) = \frac{1}{24}(9 \times 1 \times 1 + 1 \times 1 \times 3 + 0 \times 1 \times 8 + 1 \times 1 \times 6 + 1 \times 1 \times 6) = 1,$$

$$n(A_2) = \frac{1}{24}(9 \times 1 \times 1 + 1 \times 1 \times 3 + 0 \times 1 \times 8 - 1 \times 1 \times 6 - 1 \times 1 \times 6) = 0,$$

$$n(E) = \frac{1}{24}(9 \times 2 \times 1 + 1 \times 2 \times 3 - 0 \times 1 \times 8 + 1 \times 0 \times 6 + 1 \times 0 \times 6) = 1,$$

$$n(T_1) = \frac{1}{24}(9 \times 3 \times 1 - 1 \times 1 \times 3 + 0 \times 0 \times 8 - 1 \times 1 \times 6 + 1 \times 1 \times 6) = 1,$$

$$n(T_2) = \frac{1}{24}(9 \times 3 \times 1 - 1 \times 1 \times 3 + 0 \times 0 \times 8 + 1 \times 1 \times 6 - 1 \times 1 \times 6) = 1.$$

Note that the 20 crystal field terms found in this weak field approach are the same as those found in the strong field approach. The energy levels for Cr^{3+} ions are shown in Fig. 4.4 for different symmetry environments.

The optical transitions are technically forbidden transitions since they are between states of the same (even) parity. However, in the exact quantum mechanical treatment there is an admixing of the energy levels due to odd components of the crystal field and coupling with odd parity lattice vibrations. This leads to forced electric dipole transitions as discussed previously and the selection rules can be found from group theory as described in Sect. 4.3. In O_h symmetry, the electric dipole moment operator transforms as the T_{2g} irreducible representation. The

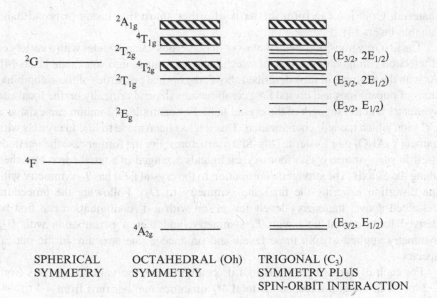

Fig. 4.4 Energy levels for a Cr^{3+} ion in different symmetry environments

Table 4.4 Forced electric dipole transitions for Cr^{3+} ions in O_h symmetry

$A_{1g} \leftrightarrow T_{2g}$
$A_{2g} \leftrightarrow T_{1g}$
$E_g \leftrightarrow T_{2g}, E_g \leftrightarrow T_{1g}$
$T_{1g} \leftrightarrow A_{2g}, T_{1g} \leftrightarrow E_g, T_{1g} \leftrightarrow T_{1g}, T_{1g} \leftrightarrow T_{2g}$
$T_{2g} \leftrightarrow A_{1g}, T_{2g} \leftrightarrow E_g, T_{2g} \leftrightarrow T_{1g}, T_{2g} \leftrightarrow T_{2g}$

transition matrix element is then expressed as the triple cross product of the irreducible representations for the initial and final states of the transition T_{2g}. The reduction of the product representation must contain the totally symmetric A_{1g} representation for the transition to be forced electric dipole allowed. For this example the allowed transitions are those listed in Table 4.4 and all other transitions are forbidden. Note that transitions between states of the same multiplicity are spin allowed while those between states of different multiplicity are spin forbidden. The lowest energy emission transition in ruby is between a ^2E_g and a $^4A_{2g}$ state. Group theory shows that this is both spin forbidden and forced electric dipole forbidden and thus has a very long fluorescent lifetime which makes it an excellent metastable state for laser emission.

In this example, group theory has been used to determine the crystal field energy levels of a Cr^{3+} ion in an octahedral symmetry environment. More detailed quantum mechanical calculations are needed to determine the magnitude of the energy level splittings. The optical absorption and emission spectra appear as a series of sharp lines and broad bands which are explained as spin-flip and configuration changing transitions [4]. Depending on the strength of the crystal field of the host material, Cr^{3+} ion can form the basis of either sharp line lasers or broad-band tunable lasers [4].

The first row transition metal ions exist in numerous valance states with a variety of d^n electron configurations, most of which have been made into solid state lasers [4]. As with the example of ruby described above, the optical properties of these combinations of dopant ions and crystal host combinations depend critically on the local site symmetry and the strength of the crystal field. For example, chromium can exist as a Cr^{4+} ion which has a d^2 configuration. This ion has been made to lase in crystals with garnet ($Y_3Al_5O_{12}$) or forsterite (Mg_2SiO_4) structures. For the former case the ion finds itself in site symmetry with four oxygen ligands arranged in a tetrahedron stretched along the S_4 axis. The strongest contribution to the crystal field has T_d symmetry with the distortion lowering the final site symmetry to D_{2d}. Following the procedure described above, the energy levels for an ion with a d^2 configuration can first be derived for a crystal field with T_d symmetry, and then a perturbation with D_{2d} symmetry applied to split these levels and produce a fine structure in the optical spectra.

For each of the two electrons in a d^2 configuration, the value of m_l runs from $+2$ to -2 in integral steps, so the total M_L quantum number runs from $+4$ to -4. The values of M_S run from $+1$ to -1. Forming all combinations of these quantum numbers that obey the Pauli exclusion principle gives five free ion terms, three singlets and two triplets. Since Hund's rule requires that the terms with highest multiplicity represent the states with lowest energies, the two terms giving rise to the optical spectroscopic properties are ^3F and ^3P. Using the character table for T_d symmetry given in Chap. 2 and the procedure described above, a ^3F term ($L=2$) will split into $^3A_2$, $^3T_2$, and $^3T_1$ crystal field terms, while the ^3P term will become a $^2T_1$ crystal field term. Then using the compatibility relations between the T_d group and its D_{2d} subgroup along with the character tables in Chap. 2, the crystal field energy levels for an ion with a d^2 configuration in a site with a crystal field having D_{2d} symmetry are $^3B_1$, $^3B_2$, ^3E, $^3A_2$, ^3E, $^3A_2$, and ^3E. The forced electric dipole selection rules can be determined as described above. The results are useful in understanding the polarized optical spectra of d^2 ions in crystals with tetrahedral symmetry.

4.5   Example: f-Electrons

Rare earth ions are another important type of dopant for optical materials [4]. Some of these are used as phosphors, lasers, amplifiers for telecommunications, and other applications. Nd^{3+} is one of the most important dopant ions for solid state lasers. Its electronic configuration consists of an inner core of filled orbitals, three optically active electrons in the partially filled 4f orbitals, two electrons in 5s orbitals, and 6 electrons in the 5p orbitals. The important optical transitions take place between 4f orbitals. The outer shell 5s and 5p electrons shield the 4f electrons from external perturbations such as crystal fields. Thus the energy levels of Nd^{3+} ions in a host crystal are found using the weak field approach. The free ion terms are found by considering three electrons each with quantum numbers $n=4$ and $l=3$ and then applying spin–orbit coupling to find the multiplet.

Considering all possible allowed combinations of the sets of m_l and m_s quantum numbers for the three electrons gives 17 free ion terms: ^2L, ^2K, ^4I, ^2I, ^2H, ^2H, ^4G, ^2G, ^2G, ^4F, ^2F, ^2F, ^4D, ^2D, ^2D, ^2P, and ^4S. Spin–orbit coupling splits each of these terms into states with total angular momentum quantum number $J=L+S, L+S-1, \ldots, L - S$. Thus for example, the ^4I term splits into four multiplets $^4I_{15/2}$, $^4I_{13/2}$, $^4I_{11/2}$, and $^4I_{9/2}$. This is the ground state term for the free ion with $^4I_{9/2}$ multiplet being the lowest energy level. The $^4F_{3/2}$ level is the first excited state above the ground state multiplets.

For rare earth ions, the crystal field perturbation is generally treated using the formalism of tensor algebra. In this treatment the expression for the crystal field in (4.26) is rewritten as

$$H_{cf} = \sum_{kq} B_{kq}^* \sum_i C_{kq}(\theta_i, \varphi_i). \tag{4.38}$$

The first sum is over the terms in the crystal field and the second sum is over the number of optically active electrons. The B_{kq}^* are the crystal field expansion parameters that depend on the crystal structure,

$$B_{kq}^* = -e^2 (-1)^q \langle r^k \rangle \sum_j \frac{Z_j C_{kq}(\hat{R}_j)}{r^{k+1}}. \tag{4.39}$$

The C_{kq} are tensor operators defined in terms of spherical harmonics as

$$C_{kq} = \left(\frac{4\pi}{2k+1} \right)^{1/2} Y_k^q(\theta, \varphi). \tag{4.40}$$

As in the example above, the first step in evaluating the crystal field matrix elements is to determine which terms in the crystal field expansion are nonzero. The 4f electronic states are represented by spherical harmonics with $l=3$. The matrix element of the crystal field operator between two of these 4f states each with odd parity must have even parity to be nonzero. Due to the orthogonality of the spherical harmonics, the Y_k^q functions in the crystal field expansion must have $k \leq (l_i + l_f) = 6$. Thus for this example the only terms in the crystal field are B_{2q}^*, B_{4q}^*, and B_{6q}^*. It is difficult to calculate the values of the B_{kq}^* parameters from crystallographic data. The usual treatment is to calculate the matrix elements of the C_{kq} tensor operators and determine the values of the B_{kq}^* by treating them as adjustable parameters in fitting theory to experimental data. For important laser crystals such as $Y_3Al_5O_{12}$: Nd^{3+} (Nd:YAG), all nine of the B_{kq}^* parameters have been determined from optical spectroscopy measurements [4].

Table 4.5 Character table for point group D_{2d}

D_{2d}	E	$2S_4$	C_2	$2C'_2$	$2\sigma_d$	Basis components
A_1	1	1	1	1	1	
A_2	1	1	1	-1	-1	
B_1	1	-1	1	1	-1	
B_2	1	-1	1	-1	1	z
E	2	0	-2	0	0	x,z
$D_{1/2}$	$2-2$	$\sqrt{2}-\sqrt{2}$	0	0	0	
$_2S$	$2-2$	$-\sqrt{2}\sqrt{2}$	0	0	0	
$\Gamma_{5/2}$	6	$\sqrt{2}$	0	0	0	$2D_{1/2}+_2S$

A YAG host crystal has a garnet structure and the Nd^{3+} ions substitute for the Y^{3+} ions. They are surrounded by tetrahedrons of oxygen ions. These tetrahedrons are slightly distorted so the exact site symmetry belongs to the D_2 point group. However the crystal field site symmetry is very close to being D_{2d} and using this group gives a good description of the crystal field energy levels and transition selection rules. The character table for D_{2d} is given in Table 2.15. This includes the double group operations and the two double-valued representations needed for states with half-integer spin angular momentum. The spin–orbit coupled free-ion multiplets will transform as either the $D_{1/2}$ or $_2S$ irreducible representations of this group. The procedure described in the previous example can be used to determine the crystal field splitting for each free ion multiplet designated by total angular momentum quantum number J. Since these have half-integer values for Nd^{3+}, time reversal degeneracy requires that each of the crystal field states be doubly degenerate. An example of a $J=5/2$ multiplet is shown in Table 4.5. The characters of the reducible $\Gamma_{5/2}$ representations can be seen by inspection (or the use of (2.10) to be reducible to two $D_{1/2}$ and one $_2S$ irreducible representations). Similarly, the $^4I_{9/2}$ free-ion ground state multiplet will split into five crystal field states. To determine how many of these states transform as $D_{1/2}$ and how many as $_2S$, the characters for the reducible representation $\Gamma_{9/2}$ must be found in the D_{2d} group. This can then be reduced in terms of the appropriate irreducible representations. For this case $\Gamma_{9/2} = 2D_{1/2}+3_2S$. Similar procedures can be carried out for all multiplets with $J=1/2$, $3/2,\ldots,15/2$ to find all of the crystal field states for Nd^{3+} in YAG. This is summarized for some of the low-lying energy levels of Nd^{3+} in garnet crystals in Fig. 4.5.

The selection rules for forced electric dipole transitions in Nd:YAG can also be found using group theory. In the D_{2d} symmetry group the x and y components of the electric dipole moment operator transform as the E representation while the z component transforms as the B_2 irreducible representation. Taking the triple product of the B_2 representation with the representations of the initial and final states of the transition gives a reducible representation that can be reduced in terms of the irreducible representations of the group. If the totally symmetric A_1 irreducible representation is found in this reduction, the transition is allowed. The results of doing this are summarized in Table 4.6 The different x, y, and z components indicate the polarization direction of the light involved in the transition.

Fig. 4.5 Energy levels of some of the low-lying multiplets of a Nd^{3+} ion in different symmetry environments

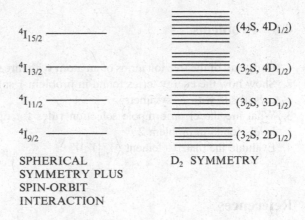

$^4F_{3/2}$ ————— ($_2$S, $D_{1/2}$)

$^4I_{15/2}$ ————— (4_2S, $4D_{1/2}$)

$^4I_{13/2}$ ————— (3_2S, $4D_{1/2}$)

$^4I_{11/2}$ ————— (3_2S, $3D_{1/2}$)

$^4I_{9/2}$ ————— (3_2S, $2D_{1/2}$)

SPHERICAL
SYMMETRY PLUS
SPIN-ORBIT
INTERACTION

D_2 SYMMETRY

Table 4.6 Selection rules for Nd^{3+} in a crystal field with D_{2d} symmetry

D_{2d}	$D_{1/2}$	$_2$S
$D_{1/2}$	x,y	x,y,z
$_2$S	x,y,z	x,y

The tensor algebra formalism mentioned above for obtaining the crystal field energy levels can be extended to the determination of strength of optical transitions. This is referred to as *Judd–Ofelt theory* [4]. The electric dipole moment operator is treated as first-rank tensor operator expressed in terms of spherical harmonic functions of order 1. The crystal field is expressed in terms of an expansion of tensor operators expressed in terms of spherical harmonics. The forced electric dipole transitions take place because the odd terms in the crystal field expansion cause an admixture of crystal field states of opposite parity. When these states of mixed parity are used in the matrix element for an electric dipole transition, the Laport selection rule no longer requires that result be zero. In this formalism the mathematical expressions for optical transition strength can be factored into a reduced matrix element that is essentially independent of crystal field symmetry and strength, and a strength parameter that reflects the configuration admixing and is sensitive to the symmetry and strength of the crystal field. These parameters are determined by fitting experimentally obtained optical spectra with theoretical expressions having adjustable parameters.

This type of analysis is helpful in understanding the optical spectroscopy properties of Nd:YAG and why it makes a good solid state laser material. All of the other first row rare earth ions existing as $+3$ or $+2$ valence states have also been made into lasers and a similar type of analysis can be used to explain their optical spectroscopic properties in different host crystals.

4.6   Problems

1. Derive all of the free ion terms of an atom with three optically active d electrons.
2. Show how the energy terms found in problem 1 split when the ion is put into a crystal site with T_d symmetry.
3. What are the electric dipole selection rules for optical transitions among the levels found in problem 2
4. Evaluate the matrix element $\left\langle Y_2^2 \middle| Y_1^1 \middle| Y_2^1 \right\rangle$.

References

1. E. Merzbacher, *Quantum Mechanics* (Wiley, New York, 1961)
2. E.U.Condon, G.H. Shortley, *The Theory of Atomic Spectra* (Cambridge University Press, London, 1935)
3. *Atomic, Molecular and Optical Physics Handbook*, ed. G.W.F. Drake (AIP, New York, 1996)
4. R.C. Powell, *Physics of Solid State Laser Materials* (Springer, New York, 1998)
5. J.S. Griffith, *The Theory of Transition Metal Ions* (Cambridge University Press, London, 1961)
6. C.J. Ballhausen, *Introduction to Ligand Field Theory* (McGraw-Hill, New York, 1962)
7. F.A. Cotton, *Chemical Applications of Group Theory* (Wiley, New York, 1963)

Chapter 5
Symmetry and the Optical Properties of Crystals

As beams of electromagnetic light waves travel through a crystal their properties may change due to their interaction with the material. In some cases these changes depend on the direction of travel in the crystal. In these cases the material is said to be optically anisotropic. The reason for this is that the structure of the crystal controls the ability of the electrons on the atoms of the crystal to respond to the influence of an electromagnetic wave. The light wave propagates through the crystal because its electric field induces the electrons on the atoms of the crystal to oscillate. If the crystal structure allows the electrons to oscillate more easily in one direction than another, then the speed of the light wave propagating in one direction will be greater than that of a light wave propagating in the other direction. This effect is called *birefringence* or double refraction. It can occur naturally due to the anisotropy of the crystal or it can be induced by an external source such as an electric field (*electrooptic effect*) or stress (*photoelastic effect*). Also the properties of the crystal may cause the light waves to exhibit *optical activity* which is a rotation of the direction of polarization. Because of the directional nature of these properties, the symmetry of the crystal plays an important role in determining the physical effects and it is possible to use transformation tensor formalism similar to that discussed in Chap. 3 to treat these optical properties. These properties have important applications in different types of light modulator devices used in a variety of optical systems. This chapter deals with "linear" optical properties while nonlinear optical effects are discussed in chap. 6

5.1 Tensor Treatment of Polarization

Due to the transverse nature of an electromagnetic light wave, the direction of its electric field vector $\mathbf{E}$ is perpendicular to the direction of propagation and is defined as the direction of polarization of the light. For unpolarized light this changes randomly with time. For polarized light the $\mathbf{E}$ vector varies in time in a regular manner. In the most general case of polarized light, the tip of the vector sweeps out an elliptical path in the distance traveled of one wavelength of the light. In special cases this reduces to a circle or a line. These are called *states of polarization* and

R.C. Powell, *Symmetry, Group Theory, and the Physical Properties of Crystals*,
Lecture Notes in Physics 824, DOI 10.1007/978-1-4419-7598-0_5,
© Springer Science+Business Media, LLC 2010

referred to as elliptically, circularly, or linearly polarized light. It is also possible to have partially polarized light that is a mixture of unpolarized light and one of the states of polarization. The physical phenomena involving the interaction of light with matter are critically dependent on the state of polarization of the light wave. Because of this, it has been useful to make optical devices that control the state of polarization of light passing through them. Some examples are linear polarizers, circular polarizers, and quarter wave plates. The first two of these devices only transmit light with states of polarization compatible to the orientation of the device. The third one retards the phase of one of the component waves with respect to the other one by $\pi/2$ which converts the state of polarization from linear to elliptical or from elliptical to linear.

The electric field of an electromagnetic light wave can always be expressed in terms of two orthogonal components. If the wave is traveling in the z direction, these components are expressed mathematically as

$$\vec{E}_x(z,t) = \hat{i}E_{0x}\cos(kz - \omega t) \tag{5.1}$$

and

$$\vec{E}_y(z,t) = \hat{j}E_{0y}\cos(kz - \omega t + \varepsilon). \tag{5.2}$$

Here k is the magnitude of the propagation vector for the wave and ε is the relative phase difference between these two components of the wave. For the case of $\varepsilon = 0$ or $\pm 2\pi$, the two component waves are "in phase" and the electric field of the light is expressed as

$$\vec{E}(z,t) = \left(\hat{i}E_{0x} + \hat{j}E_{0y}\right)\cos(kz - \omega t). \tag{5.3}$$

This describes a light wave moving in the z direction with the electric field oscillating linearly in a plane perpendicular to the z direction as shown in Fig. 5.1. If $\varepsilon = \pm\pi$, the two components are said to be "180° out of phase." The

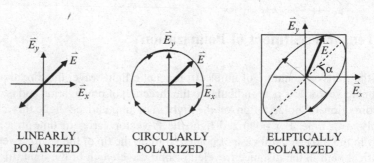

LINEARLY   CIRCULARLY   ELIPTICALLY
POLARIZED   POLARIZED   POLARIZED

Fig. 5.1 Direction of oscillation of the electric field of a light wave

electric field still oscillates linearly but in a direction shifted in position in the xy plane from the in-phase direction.

The second case of interest is circularly polarized light. In this case both constituent waves have equal amplitudes given by E_0 and their relative phase is $\varepsilon=-\pi/2$ or this amount plus any integral multiple of 2π. For this case the electric field of the wave is described by

$$\vec{E}(z,t) = E_0\left[\hat{i}\cos(kz - \omega t) + \hat{j}\sin(kz - \omega t)\right]. \qquad (5.4a)$$

For a wave coming toward an observer, (5.4a) shows that the amplitude of the electric field vector remains constant but its direction rotates clockwise with time as shown in Fig. 5.1. This is referred to as *right-circularly polarized light*. The electric field vector makes one complete revolution as the wave travels a distance of one wavelength. If the relative phase difference of the component waves is $\varepsilon=+\pi/2$ or this amount plus any integral multiple of 2π, (5.4a) becomes

$$\vec{E}(z,t) = E_0\left[\hat{i}\cos(kz - \omega t) - \hat{j}\sin(kz - \omega t)\right] \qquad (5.4b)$$

and electric field vector rotates in a counterclockwise direction. This is referred to as *left-circularly polarized light*.

In the most general case both the direction and magnitude of the electric field change as the light wave travels. This results in the tip of the electric field vector tracing out an ellipse as shown in Fig. 5.1. If the principal axes of the ellipse are aligned with the x,y coordinates, the expression for the electric field components is

$$\frac{E_x^2}{E_{0x}^2} + \frac{E_y^2}{E_{0y}^2} = 1. \qquad (5.5)$$

For this case, the orientation angle of the ellipse $\alpha=0$ or $\varepsilon=\pm\pi/2$. These are related by [1]

$$\tan 2\alpha = \frac{2E_{0x}E_{0y}\cos\varepsilon}{E_{0x}^2 - E_{0y}^2}. \qquad (5.6)$$

Natural light is generally unpolarized. Through interaction of light with matter, different states of polarization can be established. This can be achieved through reflection from a surface or transmission through different types of polarizing materials or devices.

Instead of formulating the situation in terms of light waves, it is possible to work with the irradiance of the light, $I = |E^2|$. The polarization state of any light wave can be described by four parameters known as *Stokes parameters* [1–4]

$$S_0 = 2I_0, \qquad S_1 = 2I_1 - 2I_0,$$
$$S_2 = 2I_2 - 2I_0, \quad S_3 = 2I_3 - 2I_0. \tag{5.7}$$

These are defined by considering filters that transmit only half the incident irradiance of natural light. If the first one passes all polarization states equally, the measured transmitted irradiance I_0 will be half of the incident irradiance S_0. The second filter transmits only light linearly polarized in the horizontal direction so the transmitted irradiance I_1 is equal to the incidence irradiance S_1 when the incident light is completely polarized in this direction. Similarly S_2 represents light linearly polarized at $45°$ and S_3 is a filter for circularly polarized light, opaque to linearly polarized light. Another way of stating these meanings is that: S_0 is the irradiance; S_1 is the difference between the irradiance transmitted by a linear polarizer oriented in the horizontal direction and one oriented in the vertical direction; S_2 is the difference in irradiance transmitted by a linear polarizer oriented at $+45°$ and one oriented at $135°$; and S_3 is the difference between the irradiance transmitted by a right-circular polarizer and a left-circular polarizer.

The Stokes parameters in (5.7) can be rewritten in terms of the temporal averages of the x and y electric field components as

$$S_0 = \left\langle E_{0x}^2 + E_{0y}^2 \right\rangle, \qquad S_1 = \left\langle E_{0x}^2 - E_{0y}^2 \right\rangle,$$
$$S_2 = \left\langle 2E_{0x}E_{0y} \cos \varepsilon \right\rangle, \qquad S_3 = \left\langle 2E_{0x}E_{0y} \sin \varepsilon \right\rangle. \tag{5.8}$$

Here $\varepsilon = \varepsilon_x - \varepsilon_y$, which is the relative phase difference between the two component waves and $\langle \rangle$ denotes an average over time. For monochromatic light the electric filed amplitudes and ε are all time independent and these brackets can be dropped. For unpolarized light, the x and y components of the electric field magnitude are equal so S_1, S_2, and S_3 all average to zero while S_0 does not.

The Stokes parameters can be normalized by dividing each of them by S_0. This is equivalent to assuming an incident light beam with an irradiance of one. For unpolarized light, the set of normalized Stokes parameters are $(1,0,0,0)$. The set of Stokes parameters describing linearly polarized light in the horizontal direction is $(1,1,0,0)$. For linearly polarized light in the vertical direction, the set of Stokes parameters representing the light wave is $(1, -1,0,0)$. Similar expressions can be written for other states of polarized light such as right- or left-circular polarization.

The degree of polarization for partially polarized light is defined as

$$P = \sqrt{S_1^2 + S_2^2 + S_3^2}/S_0. \tag{5.9}$$

For completely polarized light, $P = 1$ and

$$S_0^2 = S_1^2 + S_2^2 + S_3^2. \tag{5.10}$$

With this notation, the state of polarization of a light wave can then be expressed as a column vector with the four Stokes parameters as its components

$$S = \begin{pmatrix} S_0 \\ S_1 \\ S_2 \\ S_3 \end{pmatrix}. \tag{5.11}$$

S_0 is the irradiance. S_1 is the degree of linear polarization in the horizontal direction (>0) or vertical direction (<0). S_2 is the degree of linearly polarized light in the $+45°$ (>0) or $-45°$ (<0) direction. Finally, S_3 represents the degree of right-circularly polarized light (>0) or left-circularly polarized light (<0). Elliptical

Table 5.1 Polarization vectors

Polarization	Stokes vector	Jones vector
Linear		
Horizontal	$\begin{bmatrix} 1 \\ 1 \\ 0 \\ 0 \end{bmatrix}$	$\begin{bmatrix} 1 \\ 0 \end{bmatrix}$
Vertical	$\begin{bmatrix} 1 \\ -1 \\ 0 \\ 0 \end{bmatrix}$	$\begin{bmatrix} 0 \\ 1 \end{bmatrix}$
$+45°$	$\begin{bmatrix} 1 \\ 0 \\ 1 \\ 0 \end{bmatrix}$	$\frac{1}{\sqrt{2}}\begin{bmatrix} 1 \\ 1 \end{bmatrix}$
$-45°$	$\begin{bmatrix} 1 \\ 0 \\ -1 \\ 0 \end{bmatrix}$	$\frac{1}{\sqrt{2}}\begin{bmatrix} 1 \\ -1 \end{bmatrix}$
Circular		
Right	$\begin{bmatrix} 1 \\ 0 \\ 0 \\ 1 \end{bmatrix}$	$\frac{1}{\sqrt{2}}\begin{bmatrix} 1 \\ i \end{bmatrix}$
Left	$\begin{bmatrix} 1 \\ 0 \\ 0 \\ -1 \end{bmatrix}$	$\frac{1}{\sqrt{2}}\begin{bmatrix} 1 \\ -i \end{bmatrix}$
Elliptical (see text)		

polarization is the sum of linear and circular polarization. The Stokes vectors for different states of polarization are summarized in Table 5.1. This vector representation facilitates the use of a transformation tensor approach as discussed below.

A different formalism has been developed to describe the polarization state of light wave that involves a two-component column vector [1–4] This is called a *Jones vector* and is given by

$$\vec{E} = \begin{bmatrix} E_{0x}e^{i\varphi_x} \\ E_{0y}e^{i\varphi_y} \end{bmatrix}, \tag{5.12}$$

where the ϕ_i are the phases for the horizontal and vertical components of the wave. For linearly polarized light in the vertical direction, the x component of the electric field vector is zero so the Jones vector is

$$\vec{E}_v = E_{0y}e^{i\varphi} \begin{bmatrix} 0 \\ 1 \end{bmatrix}.$$

Similarly, linearly polarized light in the horizontal direction is represented by the Jones vector

$$\vec{E}_h = E_{0x}e^{i\varphi} \begin{bmatrix} 1 \\ 0 \end{bmatrix}.$$

If the amplitudes and phases of the two-component waves are the same, the Jones vector can be written as

$$\vec{E} = E_0 e^{i\varphi} \begin{bmatrix} 1 \\ 1 \end{bmatrix}.$$

This describes a light wave with linear polarization in a direction 45° between horizontal and vertical. As with the Stokes vectors, these vectors can be normalized such that the sum of the squares of the components equals one. For example,

$$\vec{E}_{45} = \frac{1}{\sqrt{2}} \begin{bmatrix} 1 \\ 1 \end{bmatrix}, \quad \vec{E}_v = \begin{bmatrix} 0 \\ 1 \end{bmatrix}, \quad \vec{E}_h = \begin{bmatrix} 1 \\ 0 \end{bmatrix}.$$

Since the relative phase difference in the two component waves for right-circularly polarized light is $-\pi/2$ and $e^{-i\pi/2} = -i$, the normalized Jones vector for this state of polarization is

$$\vec{E}_R = \frac{1}{\sqrt{2}} \begin{bmatrix} 1 \\ -i \end{bmatrix}.$$

The Jones vectors for the different common states of polarization are summarized in Table 5.1. Combinations of these vectors can be used to describe any state of polarization of a light wave.

For linearly polarized light that is not in the horizontal, vertical, or ±45° direction, the normalized Jones vector can be written as

$$E_\alpha = \begin{bmatrix} \cos\alpha \\ \sin\alpha \end{bmatrix},$$

where α is the angle between the horizontal x-axis and the direction of linear polarization. This angle of polarization is given by $\alpha = \tan^{-1}(E_{0y}/E_{0x})$.

For elliptically polarized light, the general expression for the normalized Jones vector is [4]

Orientation of ellipse	Rotation direction of ellipse
$A<B$: vertical	$C=0$; B positive, imaginary: left
$A>B$: horizontal	$C=0$; B positive, imaginary: left
$A<B$: vertical	$C=0$; B negative, imaginary: right
$A>B$: horizontal	$C=0$; B negative, imaginary: right
Between H and V	C positive; B Positive, real: left
Between H and V	C negative; B Positive, real: right

$$E_e = \frac{1}{\sqrt{A^2 + B^2 + C^2}} \begin{bmatrix} A \\ \pm B \pm iC \end{bmatrix}. \tag{5.13}$$

Note that the sum of the Jones vectors for left-circularly and right-circularly polarized light results in the Jones vector for linearly polarized light with twice the amplitude. Similarly, the sum of Jones vectors for horizontally and vertically polarized light gives the Jones vector for linearly polarized light at 45°. Thus some states of polarization can be considered to be combinations of other states of polarization. There is no Jones vector representing unpolarized light.

With the vector representation for polarization, the effects of passing a light beam through a transparent material can be described using a transformation tensor. This will be a second-rank tensor (matrix) whose dimensions depend on whether the two-component Jones vectors or four-component Stokes vectors are used to describe the state of polarization. When an optical wave with polarization described by Jones vector $\vec{E}_i$ passes through an optical material that transforms the polarization state of the transmitted wave to one described by Jones vector $\vec{E}_t$, the transformation can be described by

$$\vec{E}_t = \overset{=}{A}\vec{E}_i \qquad (5.14)$$

where $\overset{=}{A}$ is a 2×2 matrix. Written in component form this is

$$\begin{bmatrix} E_{tx} \\ E_{ty} \end{bmatrix} = \begin{bmatrix} a_{11} & a_{12} \\ a_{21} & a_{22} \end{bmatrix} \begin{bmatrix} E_{ix} \\ E_{iy} \end{bmatrix}, \qquad (5.15)$$

which can be expanded to give

$$E_{ty} = a_{21}E_{ix} + a_{22}E_{iy}.$$

Table 5.2 lists some of the *Jones matrices* that have been constructed to give specific types of polarization transitions. These can be used in conjunction with the Jones vectors from Table 5.1 to find the Jones vector describing the polarization state of a transmitted wave.

As obvious examples, light with horizontal linear polarization passing through a horizontal linear polarizer is unchanged while the same light incident on a vertical linear polarizer has no transmitted beam. These are expressed as

$$E_t = \begin{bmatrix} 1 & 0 \\ 0 & 0 \end{bmatrix} \begin{bmatrix} 1 \\ 0 \end{bmatrix} = \begin{bmatrix} 1 \\ 0 \end{bmatrix}, \quad E_t = \begin{bmatrix} 0 & 0 \\ 0 & 1 \end{bmatrix} \begin{bmatrix} 1 \\ 0 \end{bmatrix} = \begin{bmatrix} 0 \\ 0 \end{bmatrix}.$$

As a second example, consider light polarized linearly at $+45°$ incident on a quarter-wave plate with its fast axis horizontal. This is expressed as

$$E_t = e^{i\pi/4} \begin{bmatrix} 1 & 0 \\ 0 & -i \end{bmatrix} \frac{1}{\sqrt{2}} \begin{bmatrix} 1 \\ 1 \end{bmatrix} = e^{i\pi/4} \frac{1}{\sqrt{2}} \begin{bmatrix} 1 \\ -i \end{bmatrix}.$$

This transformation changes the light wave from being polarized linearly at $+45°$ to one with left-circular polarization with an additional phase factor.

The general transformation for phase retardation is [4]

$$\begin{bmatrix} e^{i\varepsilon_x} & 0 \\ 0 & e^{i\varepsilon_y} \end{bmatrix} \qquad (5.16)$$

where ε_i is the phase angle of the i-component wave. This reduces to the quarter-wave plate and half-wave plate transformation matrices shown in the table using the

Table 5.2 Polarization transformation matrices

Polarizing element	Jones matrix	Mueller matrix
Horizontal linear polarizer	$\begin{bmatrix} 1 & 0 \\ 0 & 0 \end{bmatrix}$	$\frac{1}{2}\begin{bmatrix} 1 & 1 & 0 & 0 \\ 1 & 1 & 0 & 0 \\ 0 & 0 & 0 & 0 \\ 0 & 0 & 0 & 0 \end{bmatrix}$
Vertical linear polarizer	$\begin{bmatrix} 0 & 0 \\ 0 & 1 \end{bmatrix}$	$\frac{1}{2}\begin{bmatrix} 1 & -1 & 0 & 0 \\ -1 & 1 & 0 & 0 \\ 0 & 0 & 0 & 0 \\ 0 & 0 & 0 & 0 \end{bmatrix}$
Linear polarizer at $+45°$	$\frac{1}{2}\begin{bmatrix} 1 & 1 \\ 1 & 1 \end{bmatrix}$	$\frac{1}{2}\begin{bmatrix} 1 & 0 & 1 & 0 \\ 0 & 0 & 0 & 0 \\ 1 & 0 & 1 & 0 \\ 0 & 0 & 0 & 0 \end{bmatrix}$
Linear polarizer at $-45°$	$\frac{1}{2}\begin{bmatrix} 1 & -1 \\ -1 & 1 \end{bmatrix}$	$\frac{1}{2}\begin{bmatrix} 1 & 0 & -1 & 0 \\ 0 & 0 & 0 & 0 \\ -1 & 0 & 1 & 0 \\ 0 & 0 & 0 & 0 \end{bmatrix}$
Right circular polarizer	$\frac{1}{2}\begin{bmatrix} 1 & -i \\ i & 1 \end{bmatrix}$	$\frac{1}{2}\begin{bmatrix} 1 & 0 & 0 & 1 \\ 0 & 0 & 0 & 0 \\ 0 & 0 & 0 & 0 \\ 1 & 0 & 0 & 1 \end{bmatrix}$
Left circular polarizer	$\frac{1}{2}\begin{bmatrix} 1 & i \\ -i & 1 \end{bmatrix}$	$\frac{1}{2}\begin{bmatrix} 1 & 0 & 0 & -1 \\ 0 & 0 & 0 & 0 \\ 0 & 0 & 0 & 0 \\ -1 & 0 & 0 & 1 \end{bmatrix}$
Quarter-wave plate, fast axis vertical	$e^{i\pi/4}\begin{bmatrix} -i & 0 \\ 0 & 1 \end{bmatrix}$	$\begin{bmatrix} 1 & 0 & 0 & 0 \\ 0 & 1 & 0 & 0 \\ 0 & 0 & 0 & -1 \\ 0 & 0 & 1 & 0 \end{bmatrix}$
Quarter-wave plate, fast axis horizontal	$e^{i\pi/4}\begin{bmatrix} 1 & 0 \\ 0 & -i \end{bmatrix}$	$\begin{bmatrix} 1 & 0 & 0 & 0 \\ 0 & 1 & 0 & 0 \\ 0 & 0 & 0 & 1 \\ 0 & 0 & -1 & 0 \end{bmatrix}$
Quarter-wave plate $+45°$	$\frac{1}{2}\begin{bmatrix} 1 & i \\ i & 1 \end{bmatrix}$	$\begin{bmatrix} 1 & 0 & 0 & 0 \\ 0 & 0 & 0 & -1 \\ 0 & 0 & 1 & 0 \\ 0 & 1 & 0 & 0 \end{bmatrix}$
Quarter-wave plate $-45°$	$\frac{1}{2}\begin{bmatrix} 1 & -i \\ -i & 1 \end{bmatrix}$	$\begin{bmatrix} 1 & 0 & 0 & 0 \\ 0 & 0 & 0 & 1 \\ 0 & 0 & 1 & 0 \\ 0 & -1 & 0 & 0 \end{bmatrix}$
Half-wave plate $0°$ or $90°$	$\begin{bmatrix} 1 & 0 \\ 0 & -1 \end{bmatrix}$	$\begin{bmatrix} 1 & 0 & 0 & 0 \\ 0 & 1 & 0 & 0 \\ 0 & 0 & -1 & 0 \\ 0 & 0 & 0 & -1 \end{bmatrix}$
Half-wave plate $\pm45°$	$\begin{bmatrix} 0 & 1 \\ 1 & 0 \end{bmatrix}$	$\begin{bmatrix} 1 & 0 & 0 & 0 \\ 0 & -1 & 0 & 0 \\ 0 & 0 & 1 & 0 \\ 0 & 0 & 0 & -1 \end{bmatrix}$

appropriate phase angles. The transformation matrix for rotating the direction of polarization through an angle θ is [4]

$$\begin{bmatrix} \cos\theta & -\sin\theta \\ \sin\theta & \cos\theta \end{bmatrix}. \tag{5.17}$$

A similar transformation matrix approach can be applied to polarized light described by Stokes vectors [1–3]. These *Mueller matrices* are listed in Table 5.2. They can be applied using the Stokes vectors given in Table 5.1. For example, consider an incident light beam of unit irradiance that is unpolarized. If this passes through a horizontal linear polarizer the transmitted wave is linearly polarized in the horizontal direction

$$S_t = \frac{1}{2}\begin{bmatrix} 1 & 1 & 0 & 0 \\ 1 & 1 & 0 & 0 \\ 0 & 0 & 0 & 0 \\ 0 & 0 & 0 & 0 \end{bmatrix}\begin{bmatrix} 1 \\ 0 \\ 0 \\ 0 \end{bmatrix} = \begin{bmatrix} 1/2 \\ 1/2 \\ 0 \\ 0 \end{bmatrix}.$$

The transmitted wave has an irradiance of half the irradiance of the incident wave, $S_1 > 0$, and $S_2 = S_3 = 0$. The degree of polarization as given by (5.9) is 0 for the incident wave and 1 for the transmitted wave.

As a second example, consider a partially polarized light wave described by a Stokes vector (4, 3, 0, 1). Its irradiance is 4 and its degree of polarization is about 80%. This is elliptically polarized more toward the horizontal direction ($S_1 > 0$) and right handed ($S_3 > 0$). If this goes through a quarter-wave plate oriented in the vertical direction, the result is

$$S_t = \begin{bmatrix} 1 & 0 & 0 & 0 \\ 0 & 1 & 0 & 0 \\ 0 & 0 & 0 & -1 \\ 0 & 0 & 1 & 0 \end{bmatrix}\begin{bmatrix} 4 \\ 3 \\ 0 \\ 1 \end{bmatrix} = \begin{bmatrix} 4 \\ 3 \\ -1 \\ 0 \end{bmatrix}.$$

Thus the transmitted beam has the same irradiance as the incident beam but it is now partially linearly polarized between the horizontal and $-45°$ direction with the same degree of polarization as the incident beam.

The formalism of Jones vectors and matrices is easier to use if the light is completely polarized. However, for partially polarized light it is necessary to use the Stokes vectors and Mueller matrices.

5.2 Birefringence

The dielectric properties of a crystal are expressed as

$$D_i = \varepsilon_{ij}E_j, \tag{5.18}$$

where $\mathbf{E}$ is the electric field vector, associated in this case with the optical wave, $\mathbf{D}$ is the electric displacement vector in the crystal, and $\vec{\varepsilon}$ is the dielectric constant at optical frequencies. $\mathbf{E}$ and $\mathbf{D}$ are first-rank tensors and $\vec{\varepsilon}$ is a second-rank matter tensor with nonzero components depending on the point group of the crystal as given in Table 3.4. For nonmagnetic materials, the speed of light in a crystal is

$$v = c/\sqrt{\varepsilon_{ij}} = c/n, \tag{5.19}$$

where c is the speed of light in vacuum, and at optical frequencies the refractive index is $n = \sqrt{\varepsilon_{ij}}$. For different directions of travel and polarization of the light wave in the crystal, different tensor components of the dielectric constant will be active according to (5.18), leading to different speeds of the light wave according to (5.19).

When a plane-polarized electromagnetic wave is incident on a crystal, it is bent or refracted as it travels through the material. This is due to the difference in the speed of light in the material versus its speed in free space and depends on the refractive index of the material. In the crystal the light wave can be expressed as the superposition of two plane-polarized waves with polarizations perpendicular to each other. These two waves may travel at different speeds if the refractive index of the material is anisotropic. This is called *double refraction* or *birefringence*.

For a transverse electromagnetic wave, the electric field is expressed as

$$\vec{E} = \vec{E}_0 e^{i(\vec{k}\cdot\vec{r}-\omega t)} \tag{5.20}$$

with similar expressions for $\vec{D}, \vec{B}$, and $\vec{H}$. Here $\vec{k}$ gives the direction of propagation of the light wave and has a magnitude of $(2\pi)/\lambda$, where λ is the wavelength and ω is the angular frequency of the wave. $\vec{E}_0$ gives the direction of polarization of the wave. The magnetic field vectors are perpendicular to this direction. The electric and magnetic field vectors are related to each other through Maxwell's equations. A vector $\vec{S}$ can be defined as

$$\vec{S} = \left(\frac{c}{\omega}\right)\vec{k} = n\hat{S}. \tag{5.21}$$

Note that $\vec{S} \cdot \vec{S} = n^2$. Also, $\vec{k} \times \vec{E} = -\omega\vec{B}$ and $\vec{k} \times \vec{H} = \omega\vec{D}$ which results in $\left(\frac{1}{\mu\omega^2}\right)\vec{k} \times \left(\vec{k} \times \vec{E}\right) = -\vec{\vec{D}}$.

The electric and magnetic fields expressed in the form of (5.20) along with the propagation vector of (5.21) can be used in Maxwell's equations along with appropriate vector identities to write (5.18) in tensor component form as [5]

$$\sum_j \left(\varepsilon_{ij} + S_i S_j\right) E_j = n_{ii}{}^2 E_i. \tag{5.22}$$

This is the fundamental equation for treating linear optics phenomena.

As an example of using (5.22), consider a uniaxial crystal with the symmetry axis in the x_3 direction and a wave propagating in the $x_2 x_3$ plane at an angle θ to the x_3-axis. The relevant vectors and matrices are

$$\vec{S} = \begin{pmatrix} 0 \\ n\sin\theta \\ n\cos\theta \end{pmatrix}, \quad S_i S_j = \begin{pmatrix} 0 & 0 & 0 \\ 0 & n^2 \sin^2\theta & n^2 \sin\theta\cos\theta \\ 0 & n^2 \sin\theta\cos\theta & n^2 \cos^2\theta \end{pmatrix}, \quad \overset{\leftrightarrow}{\varepsilon} = \begin{pmatrix} \varepsilon_{11} & 0 & 0 \\ 0 & \varepsilon_{11} & 0 \\ 0 & 0 & \varepsilon_{33} \end{pmatrix}.$$

Then (5.22) becomes

$$\begin{pmatrix} \varepsilon_{11} & 0 & 0 \\ 0 & \varepsilon_{11} + n^2 \sin^2\theta & n^2 \sin\theta\cos\theta \\ 0 & n^2 \sin\theta\cos\theta & \varepsilon_{33} + n^2 \cos^2\theta \end{pmatrix} \begin{pmatrix} E_1 \\ E_2 \\ E_3 \end{pmatrix} = \begin{pmatrix} n^2 & 0 & 0 \\ 0 & n^2 & 0 \\ 0 & 0 & n^2 \end{pmatrix} \begin{pmatrix} E_1 \\ E_2 \\ E_3 \end{pmatrix}.$$

Expanding this expression results in three equations

$$\varepsilon_{11} E_1 = n^2 E_1,$$
$$\left(\varepsilon_{11} + n^2 \sin^2\theta\right) E_2 + n^2 \sin\theta\cos\theta E_3 = n^2 E_2,$$
$$n^2 \sin\theta\cos\theta E_2 + \left(\varepsilon_{33} + n^2 \cos^2\theta\right) E_3 = n^2 E_3.$$

These can be solved for the two values of the index of refraction. The first equation gives the value $n_1^2 = \varepsilon_{11}$. The solution of the second two simultaneous equations is

$$\frac{1}{n_2^2} = \frac{\cos^2\theta}{n_o^2} + \frac{\sin^2\theta}{n_e^2}$$

where n_2^2 varies with θ between the values of $n_o^2 = \varepsilon_{11}$ and $n_e^2 = \varepsilon_{33}$. These results describe a uniaxial ellipsoid with axes of lengths n_o and n_e.

If the crystal is oriented with respect to the laboratory coordinate system so we can work with the principal axes as discussed in Sect. 3.2, the refractive index is represented by an ellipsoid called the *indicatrix* defined by the equation

$$\frac{x_1^2}{n_1^2} + \frac{x_2^2}{n_2^2} + \frac{x_3^2}{n_3^2} = 1, \tag{5.23}$$

where x_1, x_2, and x_3 are the principal axes. In this case the dielectric tensor is diagonal with the principal values of the dielectric constant appearing as the diagonal tensor elements. The principal values of n are related to these values through

(5.19). In this configuration, if **E** is directed along a principal axes, **D** is parallel to **E** and its magnitude depends on the principal value of ε_{ii} along that axes. If the electric field vector is not directed along a principal axes, **D** is not parallel to **E**. The details of the indicatrix and the resulting refraction depend on the symmetry of the crystal as discussed below.

For the crystal classes with cubic symmetry, T, T_h, T_d, O, and O_h, the three principal values of ε_{ii} and thus n_i are all equal. The shape of the indicatrix is spherical so all the central sections are circles. Crystals with these crystal classes do not exhibit double refraction.

Crystals having a symmetry involving one major axis of rotation such as C_4, S_4, C_{4h}, D_4, C_{4v}, D_{2d}, D_{4h}, C_3, C_{3i}, D_3, C_{3v}, D_{3d}, C_6, C_{3h}, C_{6h}, D_{6h}, D_6, C_{6v}, and D_{3h} have an indicatrix that is an ellipsoid of revolution about the principal symmetry axis. If we choose the major symmetry axis to be x_3, the indicatrix is defined by the equation

$$\frac{x_1^2}{n_o^2} + \frac{x_2^2}{n_o^2} + \frac{x_3^2}{n_e^2} = 1 \tag{5.24}$$

and is shown in Fig. 5.2. The value of the refractive index along the major symmetry axis n_e is called the extraordinary value. The value of the refractive index along each of the other two principal axes n_o is called the ordinary value. The crystal is said to be positive if $(n_e - n_o) > 0$ and it is negative if $(n_e - n_o) < 0$. The symmetry axis x_3 is called the *optic axis*. The central section perpendicular to the optic axis is a circle of radius n_o. For a wave propagating along the optic axis there is no double refraction.

The wave surfaces for ordinary and extraordinary waves for a positive uniaxial crystal are shown in Fig. 5.3. For an ordinary wave, the value of n is equal to n_o for all directions and thus the velocity of the wave is the same for all directions. For the extraordinary wave, the value of n varies with direction as n_e with its maximum

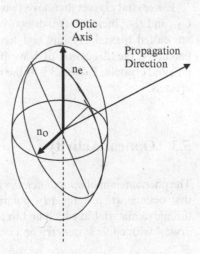

Fig. 5.2 Indicatrix

Fig. 5.3 Wave surfaces for a positive uniaxial crystal

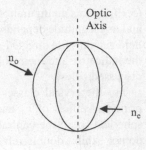

Table 5.3 Properties of some optically anisotropic materials (measured at a wavelength of 589.3 nm) [2,4]

Crystal type	Index of Refraction		
	n_o	n_e	
Uniaxial			
Calcite	1.4864	1.6584	
Quartz	1.5534	1.5443	
Sapphire	1.7681	1.7599	
Wurtzite	2.356	2.378	
Rutile	2.903	2.616	
Biaxial	n_1	n_2	n_3
Mica	1.552	1.582	1.588
Tridymite	1.469	1.47	1.473
Lanthanite	1.52	1.587	1.613
Topaz	1.619	1.620	1.627
Sulfur	1.95	2.043	2.240

value being equal to n_o along the optic axis. The velocity of this wave has the same type of variation with direction.

For crystal classes that have two major symmetry axis, C_1, C_i, C_2, C_{1h}, C_{2h}, D_2, C_{2v}, and D_{2h}, there are two directions for which no double refraction occurs. These are called biaxial crystals and have two optic axes. Except for more complex geometry, the discussion of wave propagation is the same as that given above for uniaxial crystals. Table 5.3 lists the refractive indices of several uniaxial and biaxial crystals.

5.3 Optical Activity

The phenomenon of *optical activity* refers to the rotation of the plane of polarization that occurs when a linearly polarized beam of light travels a certain distance through a material in which no birefringence occurs. This can be either an isotropic crystal with cubic symmetry or a uniaxial or biaxial crystal with the light traveling along an optic axis. The amount of rotation is determined by the thickness of the material.

A phenomenological description of optical activity can be developed by treating the light wave in the crystal as the vector sum of two component light waves, one with right-circular polarization and one with left-circular polarization. In vector form these are given by (see Table 5.1)

$$\vec{E}_{0L} = \left(E_0/\sqrt{2}\right)\begin{bmatrix} 1 \\ i \end{bmatrix}, \quad \vec{E}_{0R} = \left(E_0/\sqrt{2}\right)\begin{bmatrix} 1 \\ -i \end{bmatrix}.$$

It was shown in Sect. 5.1 that the sum of these two types of waves gives a wave with linear polarization. Due to their crystal structure, it is possible for some materials to be naturally "right handed" or "left handed" meaning that the beam with clockwise rotation travels faster or slower than the beam with counterclockwise rotation. The former is called *dextrorotatory* and the latter *levorotatory*. If this is the case, at the point where the light exits the crystal the two component beams are out of phase with each other. Since they were in phase when the light entered the crystal, their relative phase difference causes the direction of the linearly polarized $\vec{E}$ vector at the exit point to be rotated from its direction at the entrance point. For a right-handed crystal the direction of rotation is clockwise since the phase of the right-circularly polarized component increases faster than the left-circularly polarized component. For a left-handed crystal the direction of rotation is counterclockwise.

Since the direction of the linear polarization vector $\vec{E}$ bisects the direction vectors of its right- and left-circularly polarized components, the amount of rotation of the linear polarization direction is half the phase difference of its two circularly polarized components. This can be determined by the difference in the number of revolutions the right- and left-circularly polarized components make in traveling a distance d through the crystal;

$$2\pi d \left(\frac{1}{\lambda_l} - \frac{1}{\lambda_r}\right) = \frac{2\pi d}{\lambda_0}(n_l - n_r) = 2\phi. \tag{5.25}$$

Here the phase difference φ is expressed in radians and λ_0 is the wavelength of the light in vacuum. The *rotatory power* of the crystal is defined as the rotation per unit length

$$\rho = \frac{\pi}{\lambda_0}(n_l - n_r). \tag{5.26}$$

The physical source of optical activity is *spatial dispersion*, the variation of the electric field across the unit cell. This creates a contribution to the induced polarization that is proportional to the gradient of the field

$$P_i = \omega \chi_{ijl} \nabla_l E_j = i\omega \chi_{ijl} E_j k_l, \tag{5.27}$$

where the gradient of the electric field has been found using (5.20). The induced polarization is then associated with an effective dielectric tensor made up of a component from the electric field and a perturbation contribution from the gradient of the electric field

$$\varepsilon_{ij}\left(\omega, \vec{k}\right) = \varepsilon_{ij}(\omega) + i\chi_{ijl}k_l = \varepsilon_{ij}(\omega) + i\hat{e}_{ijm}g_{ml}k_l = \varepsilon_{ij}(\omega) + i\hat{e}_{ijm}G_m. \qquad (5.28)$$

In this expression, $\hat{e}_{ijm}$ is a component of an antisymmetric third-rank tensor while g_{ml} is a component of an antisymmetric second-rank tensor. In the final expression, $G_m = g_{ml}k_l$ is the *gyration tensor*. The fact that a real electric field must produce a real displacement field and to have optical activity requires that the crystal have one direction of propagation that is not equivalent to propagation in the opposite direction, places restrictions on the nonzero tensor elements as discussed below.

In the formalism developed above, a component of the electric displacement vector is expressed as

$$D_i = \varepsilon_{ij}(\omega, \vec{k})E_j = \left[\varepsilon_{ij}(\omega)E_j - i\left(\vec{\vec{G}} \times \vec{E}\right)_i\right]. \qquad (5.29)$$

The vector product in the last term is

$$\vec{\vec{G}} \times \vec{E} = \begin{pmatrix} 0 & G_{12} & G_{13} \\ -G_{12} & 0 & G_{23} \\ -G_{13} & -G_{23} & 0 \end{pmatrix} \begin{pmatrix} E_1 \\ E_2 \\ E_3 \end{pmatrix} = \begin{pmatrix} G_{12}E_2 + G_{13}E_3 \\ G_{23}E_3 - G_{12}E_1 \\ -G_{13}E_1 - G_{23}E_2 \end{pmatrix}$$

$$= \begin{pmatrix} -G_3E_2 + G_{13}E_3 \\ -G_1E_3 + G_3E_1 \\ -G_2E_1 + G_1E_2 \end{pmatrix}.$$

In the last term of this expression the relationships $-G_1=G_{23}=-G_{32}$, $-G_2= -G_{13}=G_{31}$ and $-G_3=G_{12}=-G_{21}$ have been used.

If the vector defined in (5.21) is used to designate the direction of propagation, the gyration tensor can be written as

$$\vec{\vec{G}} = G\hat{S} \qquad (5.30)$$

so the magnitude of the gyration tensor is

$$G = \hat{S}_1G_1 + \hat{S}_2G_2 + \hat{S}_3G_3. \qquad (5.31)$$

The difference in the indices of refraction in (5.26) can be defined as $G/\bar{n}$ where $\bar{n}$ is the index of refraction along the optic axis in the absence of optical activity [6]. The expression for rotatory power then becomes

$$\rho = \frac{\pi G}{\lambda_0 \bar{n}}. \tag{5.32}$$

Note that the rotatory power depends on the wavelength of light. This can lead to a rotatory dispersion effect.

Since G varies with the direction normal to the wave propagation direction, it can be expressed in terms of the direction cosines l_i of this normal direction with respect to an arbitrary axis

$$G = g_{ij} l_i l_j. \tag{5.33}$$

This is a quadratic function that can be written out as

$$G = g_{11} l_1^2 + g_{22} l_2^2 + g_{33} l_3^2 + 2 g_{23} l_2 l_3 + 2 g_{31} l_3 l_1 + 2 g_{12} l_1 l_2.$$

with components $g_{ij} = g_{ji}$. In tensor formalism

$$G = \begin{bmatrix} g_{11} & g_{12} & g_{13} \\ g_{12} & g_{22} & g_{23} \\ g_{13} & g_{23} & g_{33} \end{bmatrix}. \tag{5.34}$$

The tensor formalism discussed in Chap. 3 can be applied to G to relate the property of optical activity to the symmetry of the crystal. Since G is related to an arbitrary choice of reference axes it is important to determine how this set of axes transforms under a symmetry operation. If the set of reference axes is right handed and the transformation operation retains its right-hand nature the sign convention for the rotatory power is positive while an operation that changes the nature of the set of reference axes to left handed the sign convention is negative,

$$\rho' = \pm \rho.$$

Due to the relationship in (5.32), this same sign convention holds for G. The other transformation properties of G can be determined from (5.33). Since the l_i transform as vectors, a symmetry operation will change the components of (5.33) as follows:

$$l_i = a_{ki} l'_k, \qquad l_j = a_{mj} l'_m, \qquad G = \pm G'.$$

Thus the symmetry operation transforms (5.33) to

$$G' = g'_{km} l'_k l'_m,$$

where

$$g'_{km} = \pm a_{ki} a_{mj} g_{ij}. \tag{5.35}$$

Except for the $\pm$, this is the way the components of a second-rank tensor transform as discussed in Chap. 3. This difference in second-rank tensors is similar to the difference in polar and axial vectors (first-rank tensors).

The symmetry properties of the components of the gyration tensor can be treated using the same method described in Chap. 3 involving analyzing the subscripts. For example, an inversion operation takes $x_1 \to -x_1$, $x_2 \to -x_2$, and $x_3 \to -x_3$ so the product of any two subscripts is $+1$. However, an inversion operation changes a right-handed coordinate system to a left-handed one so the $-$ sign in (5.32) must be used. This leads to $g'_{ij} = -g_{ij}$. Since Neumann's Principle requires these tensor components to be invariant under a symmetry operation of the crystallographic point group, this result shows that a crystal with a center of symmetry cannot be optically active. As another example consider a symmetry operation of a mirror plane perpendicular to the x_2 coordinate axis. This transformation takes $x_1 \to x_1$, $x_2 \to -x_2$, and $x_3 \to x_3$ and it changes a right-handed coordinate system into a left-handed one. Thus the product of subscripts are $11 \to -11$, $12 \to 12$, $13 \to -13$, $22 \to -22$, $23 \to 23$, and $33 \to -33$. So for crystals with a point group symmetry containing this type of mirror operation, Neumann's Principle requires that the gyration tensor has the form

$$G_{\sigma_2} = \begin{bmatrix} 0 & g_{12} & 0 \\ g_{12} & 0 & g_{23} \\ 0 & g_{23} & 0 \end{bmatrix}.$$

Table 5.4 Forms of the gyration tensor crystallographic point groups

$C_{2h}, D_{2h}, C_{4h}, D_{4h},$ $S_6, D_{6h}, D_{3d}, C_{4h},$ $C_{4v}, C_{3v}, C_{6v}, C_{3h},$ D_{3h}, T, T_h, T_d, O_h	C_1	$C_2(\|x_2)$	$C_s(m \perp x_2)$
$\begin{bmatrix} 0 & 0 & 0 \\ 0 & 0 & 0 \\ 0 & 0 & 0 \end{bmatrix}$	$\begin{bmatrix} g_{11} & g_{12} & g_{13} \\ g_{12} & g_{22} & g_{23} \\ g_{13} & g_{23} & g_{33} \end{bmatrix}$	$\begin{bmatrix} g_{11} & 0 & g_{13} \\ 0 & g_{22} & 0 \\ g_{13} & 0 & g_{33} \end{bmatrix}$	$\begin{bmatrix} 0 & g_{12} & 0 \\ g_{12} & 0 & g_{23} \\ 0 & g_{23} & 0 \end{bmatrix}$
D_2	$C_{2v}, (C\|x_3)D_{2d}(C_2\|x_1)$	$C_4, D_4, C_3,$ D_3, C_6, D_6	S_4
$\begin{bmatrix} g_{11} & 0 & 0 \\ 0 & g_{22} & 0 \\ 0 & 0 & g_{33} \end{bmatrix}$	$\begin{bmatrix} 0 & g_{12} & 0 \\ g_{12} & 0 & 0 \\ 0 & 0 & 0 \end{bmatrix}$	$\begin{bmatrix} g_{11} & 0 & 0 \\ 0 & g_{11} & 0 \\ 0 & 0 & g_{33} \end{bmatrix}$	$\begin{bmatrix} g_{11} & g_{12} & 0 \\ g_{12} & -g_{11} & 0 \\ 0 & 0 & 0 \end{bmatrix}$
		T, O	
		$\begin{bmatrix} g_{11} & 0 & 0 \\ 0 & g_{11} & 0 \\ 0 & 0 & g_{11} \end{bmatrix}$	

Table 5.4 lists the form of the gyration tensor for each of the 32 crystallographic point groups. In some cases there is an arbitrary choice of orientation of a symmetry axis or mirror plane and the standard orientation is shown in the table. A different choice of orientation will change which g_{ij} components are nonzero but will not change the total number of independent components.

As an example, consider a uniaxial crystal with the symmetry axis along the x_3 direction and propagation along x_3. For this case, using (5.29) in (5.22) gives

$$\begin{pmatrix} \varepsilon_{11} & -iG_{12} & -iG_{13} \\ iG_{12} & \varepsilon_{11} & -iG_{23} \\ iG_{13} & iG_{23} & \varepsilon_{33} + n^2 \end{pmatrix} \begin{pmatrix} E_1 \\ E_2 \\ E_3 \end{pmatrix} = \begin{pmatrix} n^2 & 0 & 0 \\ 0 & n^2 & 0 \\ 0 & 0 & n^2 \end{pmatrix} \begin{pmatrix} E_1 \\ E_2 \\ E_3 \end{pmatrix},$$

where the only nonzero component of the $S_i S_j$ matrix for this propagation direction is $S_3 S_3 = n^2$. Ignoring the longitudinal (E_3) solution leaves a 2×2 determinant to be solved for the refractive index eigenvalues

$$\begin{vmatrix} \varepsilon_{11} - n^2 & -iG_{12} \\ G_{12} & \varepsilon_{11} - n^2 \end{vmatrix} = 0.$$

Expanding this gives the solutions for the two values of the refractive index as

$$n^2 = \varepsilon_{11} \pm G_{12} = \varepsilon_{11} \pm g_{33}.$$

The last term follows from $G_{12} = -G_3$ and $G_m = g_{ml}k_l$.

Rotation of the direction of polarization of an optical beam passing through a material can occur whether or not the material is naturally optically active by applying an external magnetic field in the direction of propagation. This is called the *magnetooptic effect* or *Faraday effect*. Unlike natural optical activity, which depends on the direction of optical propagation with respect to the orientation of the crystal, the Faraday effect depends only on the direction of the magnetic field with respect to the light beam. The formalism for treating magnetooptic polarization rotation is the same as described above using the gyration tensor except that the orientation direction is defined as being parallel to the applied magnetic field.

5.4 Electrooptical Effect [2, 6, 7]

At low magnitudes of the electric field E, (5.18) shows that the magnitude of the displacement D is proportional to the magnitude of the field. However, this may not be true for higher values of E. The displacement can be expressed as an expansion in terms of the magnitude of the electric field

$$D = \varepsilon E + \alpha E^2 + \cdots. \tag{5.36}$$

In general, α is negative so the second-order term lowers the value of D produced by the first term. The second term in (5.36) produces a change in the refractive index of the crystal, and this electric field-induced change in n is known as the *electrooptic effect*.

Since the dielectric constant ε is related to the refractive index, n can also be expressed as a series expansion in terms of the magnitude of the electric field

$$n = n^0 + aE + bE^2 + \cdots . \tag{5.37}$$

The electric field can be one associated with an electromagnetic wave or it can be an externally applied field, or a combination of both. In general the effect caused by an external perturbation is referred to as electrooptic while the effect caused by a strong beam of light is referred to as nonlinear optics. In this section it is assumed that an external field is present that is much larger than the field associated with a light wave so the effects of the latter contribution can be neglected. In Chap. 6, the effects of a strong electric field associated with a light wave is discussed.

In general the second terms in the expansions of (5.36) and (5.37) are the dominant terms contributing to the changes in ε or n. These contributions are called the first-order *electrooptic effect* or the *Pockels* effect. However, if we are dealing with an externally applied electric field along the major axis of symmetry, it is possible to reverse the direction of $\bar{E}$. In this case, (5.37) is the same except that the second term is negative. If the crystal has a center of symmetry, the refractive index must remain the same when the direction of the field is reversed. The only way for this to be true is for the coefficient $a=0$. Thus for crystals with a center of symmetry there is no first-order electrooptic effect. In this case there still can be a second-order electrooptic effect represented by the third term in (5.37). This is known as the *Kerr effect* and is discussed later.

The presence of an external perturbation such as an electric field introduces a preferred axis leading to anisotropic properties. Since the refractive index of a crystal is specified by the indicatrix, the small change of n produced by an electric field is essentially a small change in the size, shape, and orientation of the indicatrix. The general form of the equation for an ellipsoid is

$$\sum_{ij} B_{ij} x_i x_j = 1. \tag{5.38}$$

For an indicatrix the coefficients B_{ij} are the components of the relative dielectric impermeability tensor. Along a principal axis $B_i = 1/n_i^2$. The change in the indicatrix induced by the applied electric field can be expressed as the change in the coefficients B_{ij}

$$\Delta B_{ij} = r_{ijk} E_k, \tag{5.39}$$

where r_{ijk} are the components of the third-rank *electrooptic tensor*. Since the first-order electrooptic effect is referred to as the Pockels effect, the r_{ijk} are called the *Pockels coefficients*.

Given that $B_{ij}=B_{ji}$, the changes induced by the electric field are symmetric, $\Delta B_{ij}= \Delta B_{ji}$. This means that the coefficients of the electrooptic tensor will be symmetric with respect to the interchange of the first two subscripts, $r_{ijk}=r_{jik}$. The components of the second-rank tensor B_{ij} can be further simplified as follows

$$\begin{bmatrix} B_{11} & B_{12} & B_{31} \\ B_{12} & B_{22} & B_{23} \\ B_{31} & B_{23} & B_{33} \end{bmatrix} \Rightarrow \begin{bmatrix} B_1 & B_6 & B_5 \\ B_6 & B_2 & B_4 \\ B_5 & B_4 & B_3 \end{bmatrix}. \tag{5.40}$$

The first two subscripts of r_{ijk} are then collapsed in the same way so (5.39) becomes

$$\sum_{i=1}^{6} B_i = \sum_{j=1}^{3} r_{ij}E_j. \tag{5.41}$$

The equation for the indicatrix in the presence of an applied electric field can be written explicitly as

$$\left(\frac{1}{n^2}\right)_1 x_1^2 + \left(\frac{1}{n^2}\right)_2 x_2^2 + \left(\frac{1}{n^2}\right)_3 x_3^2 + 2\left(\frac{1}{n^2}\right)_4 x_2 x_3 + 2\left(\frac{1}{n^2}\right)_5 x_1 x_3 + 2\left(\frac{1}{n^2}\right)_6 x_1 x_2 = 1. \tag{5.42}$$

If the directions of the x_is are chosen to be along the principal directions of the dielectric tensor in the crystal when no external electric field is applied, then (5.42) must reduce to (5.28) when $\vec{E} = 0$. Then

$$\left(\frac{1}{n^2}\right)_{iE\to 0} = \frac{1}{n_i^2} \quad \text{for } i = 1, 2, 3 \quad \text{and} \quad \left(\frac{1}{n^2}\right)_{iE\to 0} = 0 \quad \text{for } i = 4, 5, 6.$$

The change in these coefficients due to an applied electric field E is written in the form of (5.41) as

$$\Delta\left(\frac{1}{n^2}\right)_i = \sum_{j=1}^{3} r_{ij}E_j \tag{5.43}$$

or in tensor form as

$$\begin{bmatrix} \Delta\left(\frac{1}{n^2}\right)_1 \\ \Delta\left(\frac{1}{n^2}\right)_2 \\ \Delta\left(\frac{1}{n^2}\right)_3 \\ \Delta\left(\frac{1}{n^2}\right)_4 \\ \Delta\left(\frac{1}{n^2}\right)_5 \\ \Delta\left(\frac{1}{n^2}\right)_6 \end{bmatrix} = \begin{bmatrix} r_{11} & r_{12} & r_{13} \\ r_{21} & r_{22} & r_{23} \\ r_{31} & r_{32} & r_{33} \\ r_{41} & r_{42} & r_{43} \\ r_{51} & r_{52} & r_{53} \\ r_{16} & r_{62} & r_{63} \end{bmatrix} \begin{bmatrix} E_1^0 \\ E_2^0 \\ E_3^0 \end{bmatrix}. \tag{5.44}$$

Table 5.5 Form of the electrooptic tensor for the crystallographic point groups

$$C_i, C_{2h}, D_{2h},$$
$$C_{4h}, D_{4h}, S_6,$$
$$D_{3d}, C_{6h}, D_{6h},$$
$$T_h, O, O_h$$

C_1

$$\begin{bmatrix} r_{11} & r_{12} & r_{13} \\ r_{21} & r_{22} & r_{23} \\ r_{31} & r_{32} & r_{33} \\ r_{41} & r_{42} & r_{43} \\ r_{51} & r_{52} & r_{53} \\ r_{61} & r_{62} & r_{63} \end{bmatrix}$$

$$\begin{bmatrix} 0 & 0 & 0 \\ 0 & 0 & 0 \\ 0 & 0 & 0 \\ 0 & 0 & 0 \\ 0 & 0 & 0 \\ 0 & 0 & 0 \end{bmatrix}$$

$C_2{}^a$

$$\begin{bmatrix} 0 & 0 & r_{13} \\ 0 & 0 & r_{23} \\ 0 & 0 & r_{33} \\ r_{41} & r_{42} & 0 \\ r_{51} & r_{52} & 0 \\ 0 & 0 & r_{63} \end{bmatrix}$$

C_{2v}

$$\begin{bmatrix} 0 & 0 & r_{13} \\ 0 & 0 & r_{23} \\ 0 & 0 & r_{33} \\ 0 & r_{42} & 0 \\ r_{51} & 0 & 0 \\ 0 & 0 & 0 \end{bmatrix}$$

$C_{1h}{}^b$

$$\begin{bmatrix} r_{11} & r_{12} & 0 \\ r_{21} & r_{22} & 0 \\ r_{31} & r_{32} & 0 \\ 0 & 0 & r_{43} \\ 0 & 0 & r_{53} \\ r_{61} & r_{62} & 0 \end{bmatrix}$$

D_2

$$\begin{bmatrix} 0 & 0 & 0 \\ 0 & 0 & 0 \\ 0 & 0 & 0 \\ r_{41} & 0 & 0 \\ 0 & r_{52} & 0 \\ 0 & 0 & r_{63} \end{bmatrix}$$

C_4

$$\begin{bmatrix} 0 & 0 & r_{13} \\ 0 & 0 & r_{13} \\ 0 & 0 & r_{33} \\ r_{41} & r_{42} & 0 \\ r_{42} & -r_{41} & 0 \\ 0 & 0 & 0 \end{bmatrix}$$

D_4

$$\begin{bmatrix} 0 & 0 & 0 \\ 0 & 0 & 0 \\ 0 & 0 & 0 \\ r_{41} & 0 & 0 \\ 0 & -r_{41} & 0 \\ 0 & 0 & 0 \end{bmatrix}$$

S_4

$$\begin{bmatrix} 0 & 0 & r_{13} \\ 0 & 0 & -r_{13} \\ 0 & 0 & 0 \\ r_{41} & -r_{51} & 0_{43} \\ r_{51} & r_{41} & 0 \\ 0 & 0 & r_{63} \end{bmatrix}$$

C_{4v}

$$\begin{bmatrix} 0 & 0 & r_{13} \\ 0 & 0 & r_{13} \\ 0 & 0 & r_{33} \\ 0 & r_{42} & 0 \\ r_{42} & 0 & 0 \\ 0 & 0 & 0 \end{bmatrix}$$

D_{2d}

$$\begin{bmatrix} 0 & 0 & 0 \\ 0 & 0 & 0 \\ 0 & 0 & 0 \\ r_{41} & 0 & 0 \\ 0 & r_{41} & 0 \\ 0 & 0 & r_{63} \end{bmatrix}$$

C_3

$$\begin{bmatrix} r_{11} & -r_{22} & r_{13} \\ -r_{11} & r_{22} & r_{13} \\ 0 & 0 & r_{33} \\ r_{41} & r_{42} & 0 \\ r_{42} & -r_{41} & 0 \\ -r_{22} & -r_{11} & 0 \end{bmatrix}$$

D_3

$$\begin{bmatrix} r_{11} & 0 & 0 \\ -r_{11} & 0 & 0 \\ 0 & 0 & 0 \\ r_{41} & 0 & 0 \\ 0 & -r_{41} & 0 \\ 0 & -r_{11} & 0 \end{bmatrix}$$

$C_{3v}{}^c$

$$\begin{bmatrix} 0 & -r_{22} & r_{13} \\ 0 & r_{22} & r_{13} \\ 0 & 0 & r_{33} \\ 0 & r_{42} & 0 \\ r_{42} & 0 & 0 \\ -r_{22} & 0 & 0 \end{bmatrix}$$

C_{3h}

$$\begin{bmatrix} r_{11} & -r_{22} & 0 \\ -r_{11} & r_{22} & 0 \\ 0 & 0 & 0 \\ 0 & 0 & 0 \\ 0 & 0 & 0 \\ -r_{22} & -r_{11} & 0 \end{bmatrix}$$

$D_{3h}{}^c$

$$\begin{bmatrix} 0 & -r_{22} & 0 \\ 0 & r_{22} & 0 \\ 0 & 0 & 0 \\ 0 & 0 & 0 \\ 0 & 0 & 0 \\ -r_{22} & 0 & 0 \end{bmatrix}$$

C_6

$$\begin{bmatrix} 0 & 0 & r_{13} \\ 0 & 0 & r_{13} \\ 0 & 0 & r_{33} \\ r_{41} & r_{42} & 0 \\ r_{42} & -r_{41} & 0 \\ 0 & 0 & 0 \end{bmatrix}$$

C_{6v}

$$\begin{bmatrix} 0 & 0 & r_{13} \\ 0 & 0 & r_{13} \\ 0 & 0 & r_{33} \\ 0 & r_{42} & 0 \\ r_{42} & 0 & 0 \\ 0 & 0 & 0 \end{bmatrix}$$

D_6

$$\begin{bmatrix} 0 & 0 & 0 \\ 0 & 0 & 0 \\ 0 & 0 & 0 \\ r_{41} & 0 & 0 \\ 0 & -r_{41} & 0 \\ 0 & 0 & 0 \end{bmatrix}$$

T, T_d

$$\begin{bmatrix} 0 & 0 & 0 \\ 0 & 0 & 0 \\ 0 & 0 & 0 \\ r_{41} & 0 & 0 \\ 0 & r_{41} & 0 \\ 0 & 0 & r_{41} \end{bmatrix}$$

[a] Parallel to x_3
[b] Perpendicular to x_3.
[c] σ_v perpendicular to x_1

Making use of Neumann's Principal and the techniques described in Chap. 3 for a third-rank tensor, the form of the electrooptic matter tensor can be determined for each of the 32 crystallographic point groups. These are given in Table 5.5. As an example, consider the crystal class C_2 which has one twofold rotation axis parallel to the z-axis. Using the normal convention for subscripts $x=1$, $y=2$, and $z=3$, this

Table 5.6 Electrooptic coefficients of several crystals

Material	Crystal class	$r\ (\times 10^{-12}\ \mathrm{m/V})$
Quartz	D_3	$r_{41}=1.4, r_{11}=0.59$
BaTiO$_2$	C_{4v}	$r_{33}=28, r_{13}=8, r_{42}=820$
LiNbO$_3$	C_{3v}	$r_{33}=30.8, r_{42}=28, r_{13}=8.6, r_{22}=3.4$
GaAs	T_d	$r_{41}=-1.5$
KDP	D_{2d}	$r_{42}=8.6, r_{63}=10.6$

180° rotation takes $1\rightarrow -1, 2 \rightarrow -2$, and $3\rightarrow 3$ and this is how the j subscript of r_{ij} transforms. Using (5.42), the i subscript of r_{ij} transforms under the C_2 operation as $1\rightarrow1, 2 \rightarrow 2, 3\rightarrow 3, 4\rightarrow -4, 5\rightarrow -5$, and $6\rightarrow 6$. Multiplying the ij subscripts shows that the only ones that remain unchanged after the symmetry operation are 13, 23, 33, 41, 42, 51, 52, and 63. This is consistent with the form of the electrooptic tensor for the C_2 point group shown in Table 5.5.

The third-rank electrooptic tensors are similar to those listed in Table 3.5 except that the difference between (3.21) and (5.44) requires a rearrangement of the tensor components. In the former equation, the third-rank tensor multiplies a second-rank tensor to give a first-rank tensor, while in the latter, the third-rank tensor multiplies a first-rank tensor to give a second-rank tensor. In the latter form no factors of 2 appear in the components of the electrooptic tensor. Table 5.6 lists the properties of several common electrooptic materials.

As an example [7], consider a material that belongs to the symmetry group D_{2d} with its fourfold symmetry axis pointed along the x_3 optic axis and the two axes with twofold symmetry pointed along x_1 and x_2. Using the appropriate electrooptic tensor from Table 5.5 in (5.44) and substituting the results into (5.42) gives the equation for the indicatrix in the presence of an electric field. If the field is applied in the x_3 direction, the equation for the indicatrix becomes

$$\frac{x_1^2}{n_o^2} + \frac{x_2^2}{n_o^2} + \frac{x_3^2}{n_e^2} + 2r_{63}E_3x_1x_2 = 1. \tag{5.45}$$

To find the standard equation for the indicatrix in the presence of the electric field it is necessary to rotate the principal axes to new directions x_1', x_2', x_3' to recover the equation for an ellipsoid

$$\frac{x_1'^2}{n_{x'1}^2} + \frac{x_2'^2}{n_{x'2}^2} + \frac{x_3'^2}{n_{x'3}^2} = 1.$$

The length of the axes of the indicatrix is given by $2n_{x_i'}$ which depends on the magnitude of the applied field. For this example, the expression in (5.45) is diagonalized by a rotation of 45° about the z-axis:

$$x_1 = x_1'\cos 45° - x_2'\sin 45°, \quad x_2 = x_1'\sin 45° + x_2'\cos 45°.$$

Using this in (5.45) gives the expression for the indicatrix referenced to the new principal axes:

$$\left[\frac{1}{n_o^2} + r_{63}E_3\right]x_1'^2 + \left[\frac{1}{n_o^2} - r_{63}E_3\right]x_2'^2 + \frac{x_3^2}{n_e^2} = 1. \tag{5.46}$$

Therefore

$$\frac{1}{n_{x_2'}^2} = \left[\frac{1}{n_o^2} - r_{63}E_3\right], \quad \frac{1}{n_{x_1'}^2} = \left[\frac{1}{n_o^2} + r_{63}E_3\right].$$

Assuming that the field-induced changes in the refractive index are small, the differential of $(1/n^2)$ can be used to solve for the new values of the refractive indices:

$$n_{x_1'} = n_o - \frac{n_o^3}{2}r_{63}E_3, \quad n_{x_2'} = n_o + \frac{n_o^3}{2}r_{63}E_3, \quad n_z = n_e.$$

Note that when the electric field goes to zero, the first two values of the refractive indices reduce to n_o as they should.

Another method to obtain this result is to make use of (5.22). For this example, (5.44) and the electrooptic tensor for D_{2d} symmetry can be used to obtain the expressions for the components of the B tensor. Since these are the inverse of the dielectric tensor components, they can be used along with the dielectric tensor for a uniaxial tensor in (5.22). The result for this example is

$$\begin{pmatrix} \varepsilon_{11} & \frac{1}{r_{63}E_3^0} & 0 \\ \frac{1}{r_{63}E_3^0} & \varepsilon_{11} & 0 \\ 0 & 0 & \varepsilon_{33} + n^2 \end{pmatrix} \begin{pmatrix} E_1 \\ E_2 \\ E_3 \end{pmatrix} = \begin{pmatrix} n^2 & 0 & 0 \\ 0 & n^2 & 0 \\ 0 & 0 & n^2 \end{pmatrix} \begin{pmatrix} E_1 \\ E_2 \\ E_3 \end{pmatrix}.$$

Expanding this gives

$$n^2 = \varepsilon_{11} - \frac{1}{r_{63}E_3^0}.$$

Since ε_{11} is the square of the initial index of refraction perpendicular to the optic axis, this expression can be rewritten as

$$\Delta\frac{1}{n^2} = -r_{63}E_3^0.$$

Expanding the differential gives the values

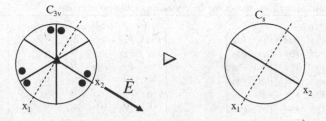

Fig. 5.4 Effect of an applied electric field on a crystal with C_{3v} symmetry

$$n_1 = n_0 \pm \frac{n_0^3}{2} r_{63} E_3^0$$

consistent with the results found above.

An interesting way to look at an external directional perturbation such as an electric field is through the symmetry stereograms shown in Fig. 1.4. The effect of the external perturbation is to remove some of the symmetry elements from the system. An example [5] of this is shown in Fig. 5.4. For this case an electric field is applied along the x_2 axis of a crystal with $3m$ symmetry. This removes all the symmetry elements except the mirror plane therefore changing the symmetry class to m. The effect is to distort the indicatrix by elongating it in the x_2 direction and decreasing its dimension in the x_1 direction.

In the example discussed above the electric field along the x_3 axis removes all the elements except E and the C_2 rotation about the z-axis from the D_{2d} symmetry group the remaining elements form the C_2 symmetry group. Thus the stereogram for D_{2d} plus E_z is equivalent for the stereogram for C_2. Comparing the electrooptic tensors for D_{2d} and C_2 in Table 5.5 shows that the only Pockels coefficients that these two tensors have in common are r_{41}, r_{52}, and r_{63}. This is consistent with the results from the analysis given above.

For materials that do not exhibit a first-order electrooptic effect, it is possible to detect a second-order electrooptic effect or Kerr effect. Since (5.39) and (5.40) show that this term in the expansion depends on the vector square of the electric field, the equation for the change in components of the indicatrix given in first order as (5.44) now becomes

$$
\begin{bmatrix}
\Delta\left(\frac{1}{n^2}\right)_1 \\
\Delta\left(\frac{1}{n^2}\right)_2 \\
\Delta\left(\frac{1}{n^2}\right)_3 \\
\Delta\left(\frac{1}{n^2}\right)_4 \\
\Delta\left(\frac{1}{n^2}\right)_5 \\
\Delta\left(\frac{1}{n^2}\right)_6
\end{bmatrix}
=
\begin{bmatrix}
\rho_{11} & \rho_{12} & \rho_{13} & \rho_{14} & \rho_{15} & \rho_{16} \\
\rho_{21} & \rho_{22} & \rho_{23} & \rho_{24} & \rho_{25} & \rho_{26} \\
\rho_{31} & \rho_{32} & \rho_{33} & \rho_{34} & \rho_{35} & \rho_{36} \\
\rho_{41} & \rho_{42} & \rho_{43} & \rho_{44} & \rho_{45} & \rho_{46} \\
\rho_{51} & \rho_{52} & \rho_{53} & \rho_{54} & \rho_{55} & \rho_{56} \\
\rho_{61} & \rho_{62} & \rho_{63} & \rho_{64} & \rho_{65} & \rho_{66}
\end{bmatrix}
\begin{bmatrix}
E_1^2 \\
E_2^2 \\
E_3^2 \\
E_2 E_3 \\
E_3 E_1 \\
E_1 E_2
\end{bmatrix}
. \tag{5.47}
$$

Here the 36 ρ_{ij} are *Kerr coefficients*. The forms for the Kerr tensor for the 32 crystallographic point groups can be found using the Neumann's Principal and

Table 5.7 Form of the Kerr tensor for the crystallographic point groups

$C_1, C_i,$

$$\begin{bmatrix} \rho_{11} & \rho_{12} & \rho_{13} & \rho_{14} & \rho_{15} & \rho_{16} \\ \rho_{21} & \rho_{22} & \rho_{23} & \rho_{24} & \rho_{25} & \rho_{26} \\ \rho_{31} & \rho_{32} & \rho_{33} & \rho_{34} & \rho_{35} & \rho_{36} \\ \rho_{41} & \rho_{42} & \rho_{43} & \rho_{44} & \rho_{45} & \rho_{46} \\ \rho_{51} & \rho_{52} & \rho_{53} & \rho_{54} & \rho_{55} & \rho_{56} \\ \rho_{61} & \rho_{62} & \rho_{63} & \rho_{64} & \rho_{65} & \rho_{66} \end{bmatrix}$$

C_2, C_{2h}, C_s

$$\begin{bmatrix} \rho_{11} & \rho_{12} & \rho_{13} & 0 & \rho_{15} & 0 \\ \rho_{21} & \rho_{22} & \rho_{23} & 0 & \rho_{25} & 0 \\ \rho_{31} & \rho_{32} & \rho_{33} & 0 & \rho_{35} & 0 \\ 0 & 0 & 0 & \rho_{44} & 0 & \rho_{46} \\ \rho_{51} & \rho_{52} & \rho_{53} & 0 & \rho_{55} & 0 \\ 0 & 0 & 0 & \rho_{64} & 0 & \rho_{66} \end{bmatrix}$$

$C_{2v}, D_2, D_{2h},$

$$\begin{bmatrix} \rho_{11} & \rho_{12} & \rho_{13} & 0 & 0 & 0 \\ \rho_{21} & \rho_{22} & \rho_{23} & 0 & 0 & 0 \\ \rho_{31} & \rho_{32} & \rho_{33} & 0 & 0 & 0 \\ 0 & 0 & 0 & \rho_{44} & 0 & 0 \\ 0 & 0 & 0 & 0 & \rho_{55} & 0 \\ 0 & 0 & 0 & 0 & 0 & \rho_{66} \end{bmatrix}$$

C_3, S_6

$$\begin{bmatrix} \rho_{11} & \rho_{12} & \rho_{13} & \rho_{14} & -\rho_{25} & \rho_{16} \\ \rho_{12} & \rho_{11} & \rho_{13} & -\rho_{14} & \rho_{25} & -\rho_{16} \\ \rho_{31} & \rho_{31} & \rho_{33} & 0 & 0 & 0 \\ \rho_{41} & -\rho_{41} & 0 & \rho_{44} & \rho_{45} & \rho_{46} \\ -\rho_{52} & \rho_{52} & 0 & -\rho_{45} & \rho_{44} & \rho_{56} \\ -\rho_{62} & \rho_{62} & 0 & \rho_{25} & \rho_{14} & \rho_{11}-\rho_{12} \end{bmatrix}$$

C_{3v}, D_3, D_{3d}

$$\begin{bmatrix} \rho_{11} & \rho_{12} & \rho_{13} & \rho_{14} & 0 & 0 \\ \rho_{12} & \rho_{11} & \rho_{13} & -\rho_{14} & 0 & 0 \\ \rho_{31} & \rho_{31} & \rho_{33} & 0 & 0 & 0 \\ \rho_{41} & -\rho_{41} & 0 & \rho_{44} & 0 & 0 \\ 0 & 0 & 0 & 0 & \rho_{44} & \rho_{56} \\ 0 & 0 & 0 & 0 & \rho_{14} & \rho_{11}-\rho_{12} \end{bmatrix}$$

C_4, C_{4h}, S_4

$$\begin{bmatrix} \rho_{11} & \rho_{12} & \rho_{13} & 0 & 0 & \rho_{16} \\ \rho_{12} & \rho_{11} & \rho_{13} & 0 & 0 & -\rho_{16} \\ \rho_{31} & \rho_{31} & \rho_{33} & 0 & 0 & 0 \\ 0 & 0 & 0 & \rho_{44} & \rho_{45} & 0 \\ 0 & 0 & 0 & -\rho_{45} & \rho_{44} & 0 \\ \rho_{61} & -\rho_{61} & 0 & 0 & 0 & \rho_{66} \end{bmatrix}$$

$C_{4v}, D_4, D_{2d}, D_{4h}$

$$\begin{bmatrix} \rho_{11} & \rho_{12} & \rho_{13} & 0 & 0 & 0 \\ \rho_{12} & \rho_{11} & \rho_{13} & 0 & 0 & 0 \\ \rho_{31} & \rho_{31} & \rho_{33} & 0 & 0 & 0 \\ 0 & 0 & 0 & \rho_{44} & 0 & 0 \\ 0 & 0 & 0 & 0 & \rho_{44} & 0 \\ 0 & 0 & 0 & 0 & 0 & \rho_{66} \end{bmatrix}$$

C_{3h}, C_6, C_{6h}

$$\begin{bmatrix} \rho_{11} & \rho_{12} & \rho_{13} & 0 & 0 & \rho_{16} \\ \rho_{12} & \rho_{11} & \rho_{13} & 0 & 0 & -\rho_{16} \\ \rho_{31} & \rho_{31} & \rho_{33} & 0 & 0 & 0 \\ 0 & 0 & 0 & \rho_{44} & \rho_{45} & 0 \\ 0 & 0 & 0 & -\rho_{45} & \rho_{44} & 0 \\ -\rho_{62} & \rho_{62} & 0 & 0 & 0 & \rho_{11}-\rho_{12} \end{bmatrix}$$

$C_{6v}, D_6, D_{3h}, D_{6h}$

$$\begin{bmatrix} \rho_{11} & \rho_{12} & \rho_{13} & 0 & 0 & 0 \\ \rho_{12} & \rho_{11} & \rho_{13} & 0 & 0 & 0 \\ \rho_{31} & \rho_{31} & \rho_{33} & 0 & 0 & 0 \\ 0 & 0 & 0 & \rho_{44} & 0 & 0 \\ 0 & 0 & 0 & 0 & \rho_{44} & 0 \\ 0 & 0 & 0 & 0 & 0 & \rho_{11}-\rho_{12} \end{bmatrix}$$

T, T_h

$$\begin{bmatrix} \rho_{11} & \rho_{12} & \rho_{13} & 0 & 0 & 0 \\ \rho_{13} & \rho_{11} & \rho_{12} & 0 & 0 & 0 \\ \rho_{12} & \rho_{13} & \rho_{11} & 0 & 0 & 0 \\ 0 & 0 & 0 & \rho_{44} & 0 & 0 \\ 0 & 0 & 0 & 0 & \rho_{44} & 0 \\ 0 & 0 & 0 & 0 & 0 & \rho_{44} \end{bmatrix}$$

T_d, O, O_h

$$\begin{bmatrix} \rho_{11} & \rho_{12} & \rho_{12} & 0 & 0 & 0 \\ \rho_{12} & \rho_{11} & \rho_{12} & 0 & 0 & 0 \\ \rho_{12} & \rho_{12} & \rho_{11} & 0 & 0 & 0 \\ 0 & 0 & 0 & \rho_{44} & 0 & 0 \\ 0 & 0 & 0 & 0 & \rho_{55} & 0 \\ 0 & 0 & 0 & 0 & 0 & \rho_{66} \end{bmatrix}$$

symmetry arguments as usual. The results are shown in Table 5.7. Note that these results are somewhat different than those given for fourth-rank tensors in Table 3.7. This is because the Kerr tensor connects two 1×6 column vectors while the eleastic compliance tensor discussed in Chap. 3 connected two 3×3 matrices.

For the Kerr coefficients ρ_{ij} both the i and j subscripts running from 1 to 6 represent xx, yy, zz, yz, xz, and xy. Thus a C_2 rotation about the z-axis leaves 20 ρ_{ij} invariant while the Kerr coefficients ρ_{14}, ρ_{15}, ρ_{24}, ρ_{25}, ρ_{34}, ρ_{35}, ρ_{41}, ρ_{42}, ρ_{43}, ρ_{46}, ρ_{51}, ρ_{52}, ρ_{53}, ρ_{56}, ρ_{64}, and ρ_{65} all must be zero. This is different from the Kerr tensor for the C_2 point group shown in Table 5.7 because the latter is derived considering a twofold rotation axis parallel to the y-axis instead of the z-axis. This demonstrates the importance of knowing the orientation of the axes when making use of these tensors in (5.37).

The external electric field that produces a linear electrooptic effect also distorts the material which introduces strain in the crystal. This is called the *piezoelectric effect* and was discussed in Chap. 3. It can be represented by a tensor with exactly the same form for each crystallographic point group as the electrooptic tensors shown in Table 5.5. In addition, there is a feedback in which the induced strain causes a change in the refractive index. This is a small perturbation of the electrooptic effect, but strain-induced changes in the refractive index can be important and are discussed in Sect. 5.5.

5.5 Photoelastic Effect

Another type of external perturbation that can change the refractive index of a crystal is stress. This is called the *photoelastic* effect. In this case, (5.37) is modified to be

$$n = n^\circ + aE_o + a'\sigma + bE_o^2 + b'\sigma^2 + \cdots, \tag{5.48}$$

where σ is the applied stress and E_o can again be either the electric field associated with a light wave in the crystal or an externally applied electric field.

The external perturbation of a stress field acts to perturb the indicatrix in the same way as the electric field discussed in Sect. 5.4. Equation (5.38) still holds but the change in the relative dielectric impermeability tensor coefficients B_{ij} given in (5.39) for the electric field now becomes

$$\Delta B_{ij} = \pi_{ijkl}\sigma_{kl} \tag{5.49}$$

for the applied stress $\vec{\vec{\sigma}}$. In this expression $\vec{\vec{\pi}}$ is a fourth-rank *photoelastic tensor* whose components are the *piezo-optical* coefficients.

Symmetry again plays a critical role in determining the π_{ijkl} coefficients that are nonzero and independent. Since $B_{ij}=B_{ji}$, the changes induced by the applied stress give $\Delta B_{ij}= \Delta B_{ji}$. Thus, $\pi_{ijkl}= \pi_{jikl}$. Also, since the stress is symmetric, $\sigma_{kl}=\sigma_{lk}$, the piezooptical coefficients are equal under the interchange of the last two subscripts, $\pi_{ijkl}= \pi_{ijlk}$. These relationships allow for the simplification of the three tensors in (5.49). The second-rank tensor for B takes the form given in (5.40) with six independent components. The expression in (5.49) then becomes

$$\Delta B_m = \sum_{n=1}^{6} \pi_{mn}\sigma_n. \tag{5.50}$$

In this expression the subscripts m and n both run from 1 to 6. The six independent components of the stress tensor can be put in a form similar to that of the B tensor in (5.40). However, as was done with the Kerr effect in the previous section, both the B and σ tensors can be expressed as 1×6 column matrices and the π tensor is a 6×6 matrix. In a way similar to the subscript rules given in Table 3.6, $\pi_{mn}=\pi_{ijlk}$ for $n=1$, 2, 3 and $\pi_{mn}=2\pi_{ijlk}$ when $n=4$, 5, 6.

Since the form of (5.47) written as tensors is the same as that of (5.44), it is not surprising that the forms of the photoelastic tensors are the same as the forms of the Kerr tensors for most of the 32 crystallographic point groups. There are three exceptions to this and they are given in Table 5.8.

As an example [6], consider the application of tensile stress along the x_1 cubic axis of a crystal with symmetry belonging to the T point group and x_2 and x_3 representing the other two cubic axes. Without the applied stress the indicatrix has a spherical shape given by

$$\frac{1}{(n^o)^2}\left(x_1^2 + x_2^2 + x_3^2\right) = 1.$$

With the applied stress the expression for the indicatrix is given by (5.42). The refractive index coefficients in this equation can be found by writing (5.50) in tensor form

$$
\begin{bmatrix}
\Delta\left(\dfrac{1}{n^2}\right)_1 \\[2ex]
\Delta\left(\dfrac{1}{n^2}\right)_2 \\[2ex]
\Delta\left(\dfrac{1}{n^2}\right)_3 \\[2ex]
\Delta\left(\dfrac{1}{n^2}\right)_4 \\[2ex]
\Delta\left(\dfrac{1}{n^2}\right)_5 \\[2ex]
\Delta\left(\dfrac{1}{n^2}\right)_6
\end{bmatrix}
=
\begin{bmatrix}
\pi_{11} & \pi_{12} & \pi_{13} & 0 & 0 & 0 \\
\pi_{13} & \pi_{11} & \pi_{12} & 0 & 0 & 0 \\
\pi_{12} & \pi_{13} & \pi_{11} & 0 & 0 & 0 \\
0 & 0 & 0 & \pi_{44} & 0 & 0 \\
0 & 0 & 0 & 0 & \pi_{44} & 0 \\
0 & 0 & 0 & 0 & 0 & \pi_{44}
\end{bmatrix}
\begin{bmatrix}
\sigma \\
0 \\
0 \\
0 \\
0 \\
0
\end{bmatrix}
=
\begin{bmatrix}
\pi_{11}\sigma \\
\pi_{13}\sigma \\
\pi_{12}\sigma \\
0 \\
0 \\
0
\end{bmatrix}.
$$

This leads to the three equations

Table 5.8 Form of the photoelastic tensors that differ from the tensor forms in Table 5.7

$$C_3, S_6$$

$$
\begin{bmatrix}
\pi_{11} & \pi_{12} & \pi_{13} & \pi_{14} & -\pi_{25} & 2\pi_{62} \\
\pi_{12} & \pi_{11} & \pi_{13} & -\pi_{14} & \pi_{25} & -\pi_{62} \\
\pi_{31} & \pi_{31} & \pi_{33} & 0 & 0 & 0 \\
\pi_{41} & -\pi_{41} & 0 & \pi_{44} & \pi_{45} & 2\pi_{52} \\
-\pi_{52} & \pi_{52} & 0 & -\pi_{45} & \pi_{44} & 2\pi_{41} \\
-\pi_{62} & \pi_{62} & 0 & \pi_{25} & \pi_{14} & \pi_{11}-\pi_{12}
\end{bmatrix}
$$

$$C_{3v}, D_3, D_{3d}$$

$$
\begin{bmatrix}
\pi_{11} & \pi_{12} & \pi_{13} & \pi_{14} & 0 & 0 \\
\pi_{12} & \pi_{11} & \pi_{13} & -\pi_{14} & 0 & 0 \\
\pi_{31} & \pi_{31} & \pi_{33} & 0 & 0 & 0 \\
\pi_{41} & -\pi_{41} & 0 & \pi_{44} & 0 & 0 \\
0 & 0 & 0 & 0 & \pi_{44} & 2\pi_{41} \\
0 & 0 & 0 & 0 & \pi_{14} & \pi_{11}-\pi_{12}
\end{bmatrix}
$$

$$C_{3h}, C_6, C_{6h}$$

$$
\begin{bmatrix}
\pi_{11} & \pi_{12} & \pi_{13} & 0 & 0 & 2\pi_{62} \\
\pi_{12} & \pi_{11} & \pi_{13} & 0 & 0 & -2\pi_{62} \\
\pi_{31} & \pi_{31} & \pi_{33} & 0 & 0 & 0 \\
0 & 0 & 0 & \pi_{44} & \pi_{45} & 0 \\
0 & 0 & 0 & -\pi_{45} & \pi_{44} & 0 \\
-\pi_{62} & \pi_{62} & 0 & 0 & 0 & \pi_{11}-\pi_{12}
\end{bmatrix}
$$

$$
\Delta\left(\frac{1}{n_1^2}\right) = \pi_{11}\sigma, \quad \Delta\left(\frac{1}{n_2^2}\right) = \pi_{13}\sigma, \quad \Delta\left(\frac{1}{n_3^2}\right) = \pi_{12}\sigma.
$$

Since B_4, B_5, and B_6 are all equal to zero, the axes of the indicatrix remain x_1, x_2, and x_3. Taking the differential of $\Delta(1/n_1^2)$ gives $-(2/n_1^3)\Delta n_1$. Substituting this into the first of these equations and solving for Δn_1 gives

$$
\Delta n_1 = -\frac{1}{2}n_1^3\pi_{11}\sigma.
$$

Since the changes in the refractive index are very small, it is a good approximation to set $n_i \cong n^o$. The three equations for the stress-induced changes in n then become

$$
\Delta n_1 = -\frac{1}{2}(n^o)^3\pi_{11}\sigma, \quad \Delta n_2 = -\frac{1}{2}(n^o)^3\pi_{13}\sigma, \quad \Delta n_3 = -\frac{1}{2}(n^o)^3\pi_{12}\sigma.
$$

The birefringence for light traveling parallel to x_2 is

$$
n_o - n_e = \Delta n_1 - \Delta n_3 = -\frac{1}{2}(n^o)^3(\pi_{11} - \pi_{12})\sigma
$$

and for light traveling parallel to x_3 it is

$$n_0 - n_e = \Delta n_1 - \Delta n_2 = -\frac{1}{2}(n^{\circ})^3(\pi_{11} - \pi_{13})\sigma.$$

An alternative way to treat the photoelastic effect is by considering the strains that the external stress causes. In Chap. 3, the strain tensor components related to the components of an external stress tensor were given by the expression

$$\varepsilon_{kl} = s_{klrs}\sigma_{rs},$$

where ε_{kl} is the strain and the s_{klrs} are components of the compliance tensor. Equation (5.49) then becomes

$$\Delta B_{ij} = p_{ijkl}\varepsilon_{kl},$$

so the *elastooptical coefficients* p_{ijkl} are given by

$$p_{ijkl} = \pi_{ijrs}(s_{rskl})^{-1}.$$

The forms of the matrices for the elastooptical tensor are essentially the same as those given in Tables 5.7 and 5.8. The factors of 2 for the matrices in Table 5.8 are not present for the elastooptical tensor. In addition, the four matrices with the π_{61} element equal to $(\pi_{11} - \pi_{12})$ have an additional factor of ½ for the p_{61} element.

5.6   Problems

1. Use the Stokes vectors and Mueller matrices to show what happens to a beam of light polarized linearly at an angle of 45° when it goes through a quarter-wave plate with its fast axis in the vertical direction. Repeat this problem using Jones vectors and Jones matrices.
2. Derive the form of the gyration tensor for a crystal with C_{2v} symmetry with the C_2 axis parallel to the x_1 direction.
3. Derive the form of the electrooptic tensor for a crystal with C_{2v} symmetry.
4. Derive the form of the Kerr tensor for a crystal with C_{2v} symmetry.

References

1. E. Hecht, *Optics* (Addison-Wesley, Reading, 1987)
2. R. Guenther, *Modern Optics* (Wiley, New York, 1990)
3. M. Born, E. Wolf, *Optics* (Pergamon, Oxford, 1965)

4. F.l. Pedrotti, L.S. Pedrotti, *Introduction to Optics* (Prentice-Hall, Englewood Cliffs, 1987)
5. A.M. Glazer, K.G. Cox, in *International Tables for Crystallography Volume D, Physical Properties of CrystalS*, ed. A. Authier (Kluwer, Dordrecht, 2003), p. 150
6. J.F. Nye, *Physical Properties of Crystals* (Oxford, London, 1957)
7. A. Yariv, *Quantum Electronics* (Wiley, New York, 1989)

Chapter 6
Nonlinear Optics

In the early 1960s the development of lasers provided light sources of sufficient power to produce nonlinear optical effects in solids. Nonlinear optics Not only has developed into a major field of research but also has found important applications in optical systems that require control and modulation of laser beams. Since lasers generally operate at a fixed wavelength or narrow range of wavelengths, one important application of nonlinear optics is to shift a laser wavelength to new wavelengths thus providing versatility necessary for many applications. This can be achieved by frequency mixing or parametric interactions. As an important example of this type of process the current chapter focuses on the nonlinear optical process of second-harmonic generation (SHG). This example demonstrates the importance of crystal structure and symmetry in these types of processes. Much of what is discussed in this chapter depends on the concepts of light beam polarization and crystal birefringence discussed in Chap. 5. It is similar to the electrooptical effect discussed in Sect. 5.4 except that the electric field causing the effect is associated with a light wave instead of an external perturbation.

There are different ways to treat the nonlinear optical properties of materials. These include nonlinear terms in the dielectric susceptibility, the refractive index, and the polarizability. They are all related since the susceptibility is a complex quantity with the imaginary part related to absorption and the real part to polarizability. The polarizability is related to the index of refraction through the Clausius–Mossotti relationship. Thus, the linear relationships are

$$\chi_{re}^{(1)} = \pi N \alpha_p, \quad n^2 = 1 + 4\pi N f_L \alpha_p.$$

Here α_p is the polarizability, N is the number of atoms or molecules, and f_L is the Lorentz local field factor. When external perturbations cause changes in χ, α_p, or n, any of these parameters can be expressed in terms of an expansion with the higher order terms representing nonlinear effects. The conventional choice of which of these parameters to use in the treatment of a specific nonlinear optical process generally depends on the cause of the nonlinear effect.

The literature for nonlinear optics makes use of equations expressed in both SI and cgs units. In general, it is common for experimental papers to utilize the

R.C. Powell, *Symmetry, Group Theory, and the Physical Properties of Crystals*, Lecture Notes in Physics 824, DOI 10.1007/978-1-4419-7598-0_6, © Springer Science+Business Media, LLC 2010

Table 6.1 Conversion between SI and cgs units

Parameter	SI	cgs	Conversion
Electric field, E	V/m	sV/cm=(erg/ cm^3)$^{1/2}$	$E(\text{cgs})=(4\pi\varepsilon_0)^{1/2}E(\text{SI})$ $E(\text{cgs})=3.33\times10^{-5}E(\text{SI})$
Second-order polarization, P	C/m^2	sC/cm^2=(erg/ cm^3)$^{1/2}$	$P(\text{cgs})=(4\pi\varepsilon_0)^{-1/2}P(\text{SI})$ $P(\text{cgs})=3\times10^5 P(\text{SI})$
Second-order susceptibility, $\chi^{(2)}$	m/V	cm^2/sC=(cm^3/ erg)$^{1/2}$	$\chi^{(2)}(\text{cgs})=\varepsilon_0(4\pi\varepsilon_0)^{-3/2}\chi^{(2)}(\text{SI})$ $\chi^{(2)}(\text{cgs})=3/(4\pi)\times10^4\chi^{(2)}(\text{SI})$
Vacuum permittivity, ε_0	mC/V	–	–
Intensity, I $I(\text{cgs})=(nc/2\pi)E^2(\text{cgs});\quad I(\text{SI})=(2\varepsilon_0 nc)E^2(\text{SI})$	W/m^2	erg/s/cm^2	$I(\text{cgs})=10^3 I(\text{SI})$
Power, S	W	erg/s	$S(\text{cgs})=10^7 S(\text{SI})$

former set of units and theoretical papers to utilize the latter. The development of nonlinear optics expressions in this chapter uses cgs units but the conversion between unit systems is listed in Table 6.1.

6.1 Basic Concepts

When a high power laser beam at a specific frequency travels through a material, the transmitted light can have both the initial frequency and new frequencies. This can provide wavelength flexibility for optical systems utilizing high power lasers. The nonlinear response can be either an elastic process such as harmonic generation that conserves optical energy, or an inelastic process such as Raman scattering in which there is an energy exchange between the light field and the medium. The most common elastic process is frequency doubling (second-harmonic generation) where part of the transmitted light has a frequency twice that of the incident light. This is the type of process that is discussed in this chapter. Inelastic processes such as Raman scattering are discussed in Chap. 7.

The standard approach to nonlinear optics is to analyze the response of a material to the electric field of an intense light beam at the atomic level. The most important effect in a dielectric material is the displacement of the valence electrons from their normal orbits creating electric dipoles resulting in macroscopic polarization of the material. At low intensities, the induced polarization is linearly proportional to the sinusoidally oscillating electric field of the incident beam, and the radiating dipoles produce an outgoing beam at the same frequency. However, at high incident intensities the oscillations of the induced dipoles may not accurately follow the frequency of the incoming wave and different frequency components are contained in the radiated wave. This is expressed as [1–3]

$$P_l(\omega_j) = \chi^{(1)}_{lm}E_m(\omega_j) + \chi^{(2)}_{lmn}E_m(\omega_r)E_n(\omega_s) + \cdots, \tag{6.1}$$

where $\mathbf{P}(\omega_i)$ is the induced polarization per unit volume and $\mathbf{E}(\omega_i)$ is the electric field vector of the light wave at frequency ω_i. The susceptibility tensor $X^{(j)}$ is of rank $j+1$ with indices l,m,n referring to Cartesian coordinates.

For low intensity light waves the magnitude of the electric field is small and the first term dominates (6.1) leading to the normal linear optics response of dielectric materials where the index of refraction and dielectric constant of the material ε are related by

$$n = (1 + 4\pi X^{(1)})^{1/2} = \varepsilon^{1/2}. \tag{6.2}$$

In SI units the factor of 4π does not appear in this expression. Since nonlinear optics is generally studied in spectral regions where the material does not absorb, the imaginary part of $X^{(1)}$ is taken to be zero. In this case the real part of $X^{(1)}$ is the polarizability tensor.

The third-rank tensor $X^{(2)}$ is responsible for optical mixing of three waves of frequency ω_i traveling in the material, resulting in sum or difference frequency generation. *Second-harmonic generation* (SHG) is a special case of sum frequency mixing with $\omega_3 = \omega_1 + \omega_2$. Since $\omega_1 = \omega_2$, in this case $\omega_3 = 2\omega_1$. From a photon perspective, this is conservation of energy as shown in Fig. 6.1. The two incoming photons make transitions to virtual electronic states of the material and the outgoing photon is emitted from the higher virtual state. The polarization term for SHG becomes

$$P_l(2\omega) = X^{(2)}{}_{lmn} E_m(\omega) E_n(\omega). \tag{6.3}$$

To use this three wave mixing formalism for SHG, the first two waves are considered to be two orthogonally polarized components of the incident laser beam. Each of these components has a frequency ω_1 and they create a polarization

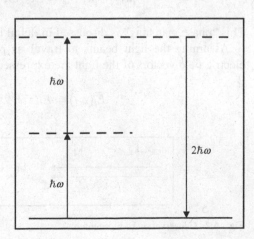

Fig. 6.1 Photon transitions in second-harmonic generation

wave that has a component at frequency $\omega_3 = 2\omega_1$. The incident wave has a wavelength and phase velocity in the material given by

$$\lambda_1 = \frac{c}{v_1 n_1}, \quad v_1 = \frac{c}{n_1} \tag{6.4}$$

Energy is transferred from the polarization wave to a light wave traveling at frequency $2\omega_1$ with a phase velocity and wavelength given by

$$\lambda_3 = \frac{c}{v_3 n_3}, \quad v_3 = \frac{c}{n_3}. \tag{6.5}$$

For efficient energy transfer to occur, the polarization wave and the second-harmonic wave must remain in phase. This will only occur if they have the same phase velocity which requires that $n_3 = n_1$. The waves travel with wave vectors having magnitudes k_i that obey the dispersion relation

$$k_i(\omega_i) = \omega_i n(\omega_i)/c = 2\pi n(\omega_i)/\lambda_i.$$

In regions of normal dispersion (far from regions of absorption), the phase mismatch given in wavenumbers for collinear wave vectors is

$$\Delta k = 2\pi \left(\frac{2n_3}{\lambda_1} - \frac{n_1}{\lambda_1} - \frac{n_1}{\lambda_1} \right) = \frac{4\pi}{\lambda_1}(n_3 - n_1). \tag{6.6}$$

Since k_i is a vector, conservation of momentum can be satisfied by co-linear waves or the vector conditions shown in Fig. 6.2. For SHG where $\mathbf{k}_1$ and $\mathbf{k}_2$ are components of the same beam of light the co-linear case applies. For the fundamental and second-harmonic waves to remain in phase as they travel through the crystal

$$|\mathbf{k}_1| = |\mathbf{k}_2|, k_3 = 2k_1, \omega_3 = 2\omega_1, n_3 = n_1.$$

This "phase matching" is discussed in detail below.

Assuming the light beams to travel as plane waves, the components of the electric field vectors of the light are expressed as

$$\vec{E}_i(z, t) = \vec{E}_i(z)e^{i(k_i z - \omega t)} + c.c. \tag{6.7}$$

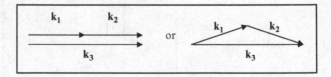

Fig. 6.2 Conservation of momentum.

For propagation in the z direction Maxwell's wave equation is [4]

$$\frac{\partial^2 E}{\partial z^2} = \frac{1}{c^2}\frac{\partial^2 E}{\partial t^2} + \frac{4\pi}{c^2}\frac{\partial^2 P_{NL}\left(=\chi^{(2)}\vec{E}\vec{E}\right)}{\partial t^2}. \tag{6.8}$$

Note that in SI units the 4π in the last term is replaced by ε_0^{-1}. Substituting the expression for the electric fields on the right-hand side of this equation, using $\omega_3 = \omega_1 + \omega_2 = 2\omega_1$, and the slowly varying envelope approximation

$$\left|\frac{d^2 E}{dz^2}\right| \ll \left|k\frac{dE}{dz}\right|$$

gives

$$\frac{dE_3(z)}{dz} = \frac{i2\pi\omega_3^2}{c^2 k_3}\chi^{(2)}E_1(z)E_2(z)e^{-i\Delta kz}. \tag{6.9}$$

Here Δk is given by (6.6) and the polarization vector directions of the electric fields are temporarily being neglected. Similar equations are obtained for E_1 and E_2. This "coupled wave equation" describes the increase in the wave amplitude represented by E_3 due to the nonlinear interaction of waves E_1 and E_2 over a distance z in the material. The expressions for E_1 and E_2 represent back transfer from E_3 to the original waves.

Integrating both sides of (6.9) from zero to a distance L in the material gives

$$E_3(L) = \frac{2\pi i \chi^{(2)}\omega_3^2}{c^2 k_3}E_1(L)E_2(L)\left[\frac{e^{-i\Delta kL} - 1}{\Delta k}\right].$$

In deriving this expression it has been assumed that there is no beam depletion so $E_1(z)$ and $E_2(z)$ were removed from the integral over z.

The intensity of the second-harmonic beam of light can be found from the square of the electric field with the appropriate factors given in Table 6.1:

$$I_3(L) = \frac{n_3 c}{2\pi}E_3^2 = \frac{n_3}{2\pi}\frac{(2\pi)^2\left[\chi^{(2)}\right]^2\omega_3^4}{c^3 k_3^2}E_1^2(L)E_2^2(L)\left[\frac{e^{-i\Delta kL} - 1}{\Delta k}\right]^2.$$

The next steps are to square the final factor in this expression, convert the electric filed amplitudes to intensities using the factors from Table 6.1, and express the factors of ω_3 and k_3 in terms of the wavelength λ_3. This gives

$$I_3 = \frac{512\pi^5 d_{eff}^2 L^2}{cn_1^2 n_3 \lambda_1^2}I_1 I_2 \frac{\sin^2\left(\Delta kL/2\right)}{\left(\Delta kL/2\right)^2}. \tag{6.10}$$

Here L is the length traveled in the material and all of the numerical factors have been combined. In addition, the facts that $n_1=n_2$ for SHG and $\lambda_3=\lambda_1/2$ have been used. d_{eff} is the effective nonlinear coefficient. Its magnitude is defined in terms of the nonlinear polarizability tensor as $\mathbf{X}_{\text{eff}}^{(2)}/2$. At this point the vector polarization directions for the electric fields are put back into the equation. The effective nonlinear optic coefficient is a third-rank tensor that is defined in terms of the unit polarization vectors of the incident light beam and the SHG beam as

$$d_{\textit{eff}} = \hat{p}_3 \cdot \overset{=}{d} \hat{p}_1 \hat{p}_2. \tag{6.11}$$

This will be discussed further below.

The conversion efficiency for SHG is found by dividing both sides of (6.10) by I_1. The power of the SHG beam is found by multiplying the expression in (6.10) by the cross-sectional area of the beam. The expression in (6.10) was derived using the assumption of no depletion of the incident light beam. For many cases this is a good approximation. If incident beam depletion is significant the expression for the SHG beam intensity becomes more complicated [2]. However, most of the fundamental topics discussed below remain the same.

As seen from (6.10), second-harmonic generation depends critically on the phase mismatch Δk. The functional dependence of SHG conversion on phase mismatch is shown in Fig. 6.3. It is highly peaked at $\Delta k=0$ where there is exact phase matching. The SHG intensity increases and decreases with the distance traveled in the crystal L with a period of $\Delta kL/2=\pi$. The SHG is maximum at half of this distance as measured from the front of the crystal. This distance is called the *coherence length*, l_c. Using (6.6) it is given by

$$l_c = \frac{\lambda_1}{4(n_3 - n_1)}. \tag{6.12}$$

For perfect phase matching $n_3=n_1$ so $l_c=\infty$.

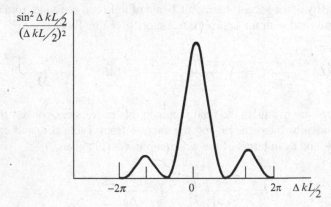

Fig. 6.3 Dependence of SHG power on phase mismatch

The final factor in (6.10) can be expressed in terms of coherence length so the intensity of the SHG beam is

$$I_3 \propto \sin^2\left(\frac{\pi L}{2l_c}\right) \qquad (6.13)$$

Note that this intensity oscillates with propagation distance in the crystal with a period of $L=2l_c$. This is because the coupled electromagnetic waves can transfer energy in both directions and the decrease in SHG power is due to transfer of energy from the SHG wave back to the fundamental wave. For perfect phase matching the SHG intensity is given by

$$I_3 = \frac{512\pi^5 d_{eff}^2 L^2}{c n_1^3 \lambda_1^2} I_1^2. \qquad (6.14)$$

The intensity of the frequency doubled beam is proportional to the square of the length of the crystal.

The above discussion demonstrates the importance of matching the indices of refraction of the incident light wave and the frequency doubled light wave. In general the index of refraction of a material exhibits a normal dispersion with frequency (far from any region of absorption). This varies with temperature and with the direction of beam propagation and polarization in the crystal. Controlling these parameters to minimize phase mismatch in nonlinear optics is referred to as "temperature-tuned phase matching" and "angle-tuned phase matching". The latter will be discussed below. Proper angle tuning provides a spatial resonance of the interaction of the light waves. In isotropic crystals it is always true that $n_1 > n_3$ so no phase matching can occur. Thus it is necessary to have an anisotropic crystal with waves having different polarizations. This makes it possible to utilize the birefringence properties described in Chap. 5 in order to achieve phase matching.

As an example, consider a plane wave incident on a uniaxial crystal with its optic axis parallel to the z direction. The "principal plane" is defined as the one containing the vectors **k** and **z**. **E** polarized perpendicular to the principal plane is called an "ordinary wave" and **E** polarized in the principal plane is called an "extraordinary wave." This configuration is shown in Fig. 6.4. The polarization direction for an ordinary wave is always perpendicular to **z** so the value of its refractive index, n^o, is independent of the direction of **k**. However, the polarization direction for an extraordinary wave varies from being parallel to being perpendicular to **z** depending on the direction of **k**. Thus the value of the refractive index for an extraordinary wave, n^e, varies with the angle θ between **k** and **z**. The difference in values of these two indices of refraction

$$\Delta n = n^o - n^e(\theta) \qquad (6.15)$$

Fig. 6.4 Propagation
directions in a uniaxial crystal

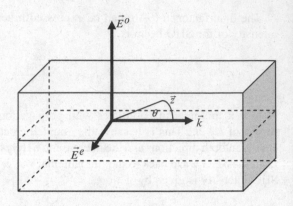

is called the "birefringence." $\Delta n = 0$ for **k** along **z** and is maximum for **k** perpendicular to **z**. The "principal value" of the refractive index of the extraordinary wave is designated n_e which is the value when $\theta = 90°$. Since the value of the refractive index of the ordinary wave is independent of θ, this is its principal value n_o. An expression for the dependence of $n^e(\theta)$ on the angle θ can be derived by breaking up the phase velocity of the extraordinary wave into its components parallel and perpendicular to the z axis, $\vec{v}_p^e(\theta) = \vec{v}_p(\perp z) + \vec{v}_p(\|z)$. Since $v = c/n$ and this is a right triangle, the magnitudes of these vectors are related by

$$\left(\frac{c}{n^e(\theta)} \right)^2 = \left(\frac{c \sin \theta}{n_e} \right)^2 + \left(\frac{c \cos \theta}{n_o} \right)^2$$

Using appropriate trigonometry identities this can be solved for $n^e(\theta)$ to give

$$n^e(\theta) = n_o \left[\frac{(1 + \tan^2 \theta)}{\left(1 + \left({n_o}/{n_e} \right)^2 \tan^2 \theta \right)} \right]^{1/2} \tag{6.16}$$

With the coordinates shown in Fig. 6.5

$$
\begin{aligned}
n^o(\theta) &= n_o & \Delta n(\theta) &= n_o - n^e(\theta) \\
n^e(0) &= n_o & \Delta n(0) &= 0 \\
n^e(90°) &= n_e & \Delta n(90°) &= n_o - n_e.
\end{aligned}
\tag{6.17}
$$

If $n_o > n_e$ the crystal is said to be "negative." If $n_o < n_e$ the crystal is said to be "positive." The geometry of birefringent phase matching is discussed further in Sect. 6.3.

From this discussion of the basic concepts of frequency doubling, it is clear that the critical factors for achieving efficient second-harmonic generation are

Fig. 6.5 Coordinate system **k**

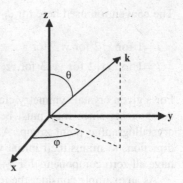

maximizing the effective nonlinear optical coefficient and minimizing the phase mismatch between the fundamental and second-harmonic light wave. Methods for achieving these criteria are discussed below.

6.2 Effective Nonlinear Optical Coefficient

In addition to phase matching, (6.10) shows that the effective nonlinear optical coefficient is a critical parameter in determining the efficiency of second-harmonic generation. Since $X^{(2)}=2d$, the second term in (6.1) can be rewritten as

$$P_i^{NL} = 2d_{il}E_l^2. \tag{6.18}$$

Thus $\vec{\vec{d}}$ is a third-rank matter tensor connecting a cause represented by a vector squared with an effect represented by a vector. The polarization wave in (6.18) is the origin of the second-harmonic light wave.

In general a third-rank tensor has 27 components but crystal symmetry limits the number of unique, nonzero components as discussed in Chap. 3. For practical purposes, crystals used for SHG generally have one dominant coefficient that maximizes the effect for a specific direction of light propagation. The components of the nonlinear optical tensor d_{ij} are determined through symmetry, and d_{eff} defined in (6.11) is derived based on the polarization vectors of the interacting waves.

The tensor component expression for (6.18) is

$$
\begin{pmatrix} p_x \\ p_y \\ p_z \end{pmatrix} = \begin{pmatrix} d_{11}d_{12}d_{13}d_{14}d_{15}d_{16} \\ d_{21}d_{22}d_{23}d_{24}d_{25}d_{26} \\ d_{31}d_{32}d_{33}d_{34}d_{35}d_{36} \end{pmatrix} \begin{pmatrix} E_x^2 \\ E_y^2 \\ E_z^2 \\ 2E_yE_z \\ 2E_xE_z \\ 2E_xE_y \end{pmatrix}. \tag{6.19}
$$

The convention used here for d_{il} is the same as that used previously:

$i = 1$ for x; 2 for y; 3 for z

$l = 1$ for xx; 2 for yy; 3 for zz; 4 for $yz = zy$; 5 for $xz = zx$; and 6 for $xy = yx$.

For a given crystal symmetry class many of the coefficients are 0 or equal to other components since they must be invariant under all symmetry operations of the crystallographic point group. An inversion process changes all three coordinate directions to minus their initial value and thus crystals with a center of symmetry have all zero components for d_{ij} and thus SHG is not possible.

As an example consider the tetragonal symmetry group C_{4v}. This has a fourfold rotation axis about z as shown in Fig. 6.6. Under this operation, the components of d_{il} transform as

$$d_{11} = d_{xxx} \rightarrow d_{yyy} = d_{22} \ but \ d_{22} = d_{yyy} \rightarrow -d_{xxx} = -d_{11} \ \therefore \ d_{11} = d_{22} = 0$$
$$d_{12} = d_{xyy} \rightarrow d_{yxx} = d_{21} \ but \ d_{21} = d_{yxx} \rightarrow -d_{xyy} = -d_{12} \ \therefore \ d_{12} = d_{21} = 0$$
$$d_{13} = d_{xzz} \rightarrow d_{yzz} = d_{23} \ but \ d_{23} = d_{yzz} \rightarrow -d_{xzz} = -d_{13} \ \therefore \ d_{13} = d_{23} = 0$$
$$d_{14} = d_{xyz} \rightarrow -d_{yxz} = -d_{25} \ but \ d_{25} = d_{yxz} \rightarrow -d_{xyz} = -d_{14} \ \therefore \ d_{14} = d_{25} = 0$$
$$d_{15} = d_{xxz} \rightarrow d_{yyz} = d_{24} \ and \ d_{24} = d_{yyz} \rightarrow -d_{xxz} = d_{15} \ \therefore \ d_{15} = d_{24}$$
$$d_{16} = d_{xxy} \rightarrow -d_{yyx} = -d_{26} \ but \ d_{26} = d_{yxy} \rightarrow d_{xyx} = d_{16} \ \therefore \ d_{16} = d_{26} = 0$$
$$d_{31} = d_{zxx} \rightarrow d_{zyy} = d_{32} \ and \ d_{32} = d_{zyy} \rightarrow d_{zxx} = d_{31} \ \therefore \ d_{31} = d_{32}$$
$$d_{33} = d_{zzz} \rightarrow d_{zzz} = d_{33}$$
$$d_{34} = d_{zyz} \rightarrow -d_{zxz} = -d_{35} \ but \ d_{35} = d_{zxz} \rightarrow d_{zyz} = d_{34} \ \therefore \ d_{34} = d_{35} = 0$$
$$d_{36} = d_{zxy} \rightarrow -d_{zyx} = -d_{36} \ \therefore \ d_{36} = 0.$$

Thus the nonlinear optical tensor is

$$\vec{\vec{d}} = \begin{pmatrix} 0 & 0 & 0 & 0 & d_{15} & 0 \\ 0 & 0 & 0 & d_{15} & 0 & 0 \\ d_{31} & d_{31} & d_{33} & 0 & 0 & 0 \end{pmatrix}. \tag{6.20}$$

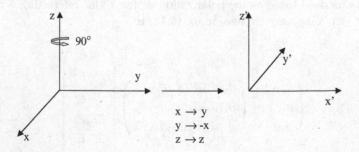

Fig. 6.6 Rotation of 90° about the z-axis

In addition, if anomalous dispersion can be ignored (which is approximately true if the frequencies of the light beams are far from absorption transitions so $\chi^{(2)}(\omega_3,\omega_1,\omega_2)$ is real), Kleinman showed that the susceptibility remains unchanged when the frequencies of the three beams are permuted. This leads to the fact that the indices of d_{il} must be invariant to all permutations [3]. For example,

$$d_{21} = d_{yxx} = d_{xxy} = d_{16}.$$

This *Kleinman symmetry* requires for all crystal classes that

$$
\begin{aligned}
d_{21} &= d_{16} & d_{13} &= d_{35} \\
d_{24} &= d_{32} & d_{14} &= d_{36} = d_{25} \\
d_{31} &= d_{15} & d_{12} &= d_{26}.
\end{aligned}
\tag{6.21}
$$

The forms of the nonlinear optical tensor for the 32 crystallographic point groups including Kleinman symmetry are given in Table 6.2. Note there is some arbitrariness as to the designation a coefficient when there are several equivalent ones. For example, if d_{15} is equivalent to d_{31} either one can be used to designate these two coefficients. This creates some confusion in the literature.

The expression for the effective nonlinear optical coefficients for specific polarization directions of the incident and frequency doubled light beams is given by (6.11). The vector product in the last factor is a six-component column tensor that can be expressed as

$$
\mathbf{p}_1 \mathbf{p}_2 =
\begin{pmatrix}
p_{1x}p_{2x} \\
p_{1y}p_{2y} \\
p_{1z}p_{2z} \\
p_{1y}p_{2z} + p_{1z}p_{2y} \\
p_{1x}p_{2z} + p_{1z}p_{2x} \\
p_{1x}p_{2y} + p_{1y}p_{2x}
\end{pmatrix}.
\tag{6.22}
$$

Any linearly polarized wave in a uniaxial crystal can be represented as a superposition of two waves with o and e polarizations. A unit vector for polarization can be defined as $|p| = 1$ with components referenced to the coordinate system shown in Fig. 6.7 given by

$$
\begin{aligned}
p_{ox} &= -\sin\phi & p_x^e &= \cos\theta\cos\phi \\
p_{oy} &= \cos\phi & p_y^e &= \cos\theta\sin\phi \\
p_{oz} &= 0 & p_z^e &= -\sin\theta.
\end{aligned}
\tag{6.23}
$$

Table 6.2 Nonlinear optical tensors for the crystallographic point groups (with Kleinmann symmetry conditions)

$C_i, C_{2h}, D_{2h}, C_{4h},$
$D_4, D_{4h}, C_{3i}, D_{3d}, C_{6h},$
D_6, D_{6h}, T_h, O, O_h C_1

$$
\begin{pmatrix} 0 & 0 & 0 & 0 & 0 & 0 \\ 0 & 0 & 0 & 0 & 0 & 0 \\ 0 & 0 & 0 & 0 & 0 & 0 \end{pmatrix}
\qquad
\begin{pmatrix} d_{11} & d_{12} & d_{13} & d_{14} & d_{15} & d_{16} \\ d_{16} & d_{22} & d_{23} & d_{24} & d_{14} & d_{12} \\ d_{15} & d_{24} & d_{33} & d_{23} & d_{13} & d_{14} \end{pmatrix}
$$

C_2 C_{1h} D_2, D_{2d}, T, T_d

$$
\begin{pmatrix} 0 & 0 & d_{13} & d_{14} & 0 & d_{16} \\ d_{16} & d_{22} & d_{23} & 0 & d_{14} & 0 \\ 0 & 0 & 0 & d_{23} & 0 & d_{14} \end{pmatrix}
\begin{pmatrix} d_{11} & d_{12} & d_{13} & 0 & d_{15} & 0 \\ 0 & 0 & 0 & d_{24} & 0 & d_{12} \\ d_{15} & d_{24} & d_{33} & 0 & d_{13} & 0 \end{pmatrix}
\begin{pmatrix} 0 & 0 & 0 & d_{14} & 0 & 0 \\ 0 & 0 & 0 & 0 & d_{14} & 0 \\ 0 & 0 & 0 & 0 & 0 & d_{14} \end{pmatrix}
$$

C_{2v} S_4

$$
\begin{pmatrix} 0 & 0 & 0 & 0 & d_{15} & 0 \\ 0 & 0 & 0 & d_{24} & 0 & 0 \\ d_{15} & d_{24} & d_{33} & 0 & 0 & 0 \end{pmatrix}
\qquad
\begin{pmatrix} 0 & 0 & 0 & d_{14} & d_{15} & 0 \\ 0 & 0 & 0 & -d_{15} & d_{14} & 0 \\ d_{15} & -d_{15} & 0 & 0 & 0 & d_{14} \end{pmatrix}
$$

C_6, C_4, C_{4v}, C_{6v} C_3 C_{3v}

$$
\begin{pmatrix} 0 & 0 & 0 & 0 & d_{15} & 0 \\ 0 & 0 & 0 & d_{15} & 0 & 0 \\ d_{15} & d_{15} & d_{33} & 0 & 0 & 0 \end{pmatrix}
\begin{pmatrix} d_{11} & -d_{11} & 0 & 0 & d_{15} & -d_{16} \\ -d_{16} & d_{16} & 0 & d_{15} & 0 & -d_{11} \\ d_{15} & d_{15} & d_{33} & 0 & 0 & 0 \end{pmatrix}
\begin{pmatrix} 0 & 0 & 0 & 0 & d_{15} & -d_{16} \\ -d_{16} & d_{16} & 0 & d_{15} & 0 & 0 \\ d_{15} & d_{15} & d_{33} & 0 & 0 & 0 \end{pmatrix}
$$

D_3 C_{3h} D_{3h}

$$
\begin{pmatrix} d_{11} & -d_{11} & 0 & 0 & 0 & 0 \\ 0 & 0 & 0 & 0 & 0 & -d_{11} \\ 0 & 0 & 0 & 0 & 0 & 0 \end{pmatrix}
\begin{pmatrix} d_{11} & -d_{11} & 0 & 0 & 0 & -d_{16} \\ -d_{16} & d_{16} & 0 & 0 & 0 & -d_{11} \\ 0 & 0 & 0 & 0 & 0 & 0 \end{pmatrix}
\begin{pmatrix} 0 & 0 & 0 & 0 & 0 & -d_{16} \\ -d_{16} & d_{16} & 0 & 0 & 0 & 0 \\ 0 & 0 & 0 & 0 & 0 & 0 \end{pmatrix}
$$

Fig. 6.7 Coordinate system for ordinary and extraordinary polarization vectors

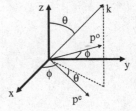

Here the Z coordinate axis is the direction of the optic axis. $\mathbf{p}^o$ is in the XY plane perpendicular to both $\mathbf{k}$ and the optic axis with its direction varying with the azimuthal angle ϕ. $\mathbf{p}^e$ is in the kZ plane with its direction varying with both θ and ϕ.

Consider as an example the case where $\mathbf{p}_1$ and $\mathbf{p}_3$ are ordinary waves and $\mathbf{p}_2$ is extraordinary. Then for a tetragonal crystal with C_{4v} symmetry, the d tensor given in Table 6.1 and the $\mathbf{p}_i$ vectors given in (6.23) can be substituted into (6.19) to give

$$d_{\text{eff}} = (-\sin\phi, \cos\phi, 0) \begin{pmatrix} 0 & 0 & 0 & 0 & d_{15} & 0 \\ 0 & 0 & 0 & d_{15} & 0 & 0 \\ d_{15} & d_{15} & d_{33} & 0 & 0 & 0 \end{pmatrix} \begin{pmatrix} -\sin\phi\cos\phi\cos\theta \\ \cos\phi\sin\phi\cos\theta \\ 0 \\ -\cos\phi\sin\theta \\ \sin\phi\sin\theta \\ (\cos^2\phi - \sin^2\phi)\cos\theta \end{pmatrix}$$

$$= (-\sin\phi, \cos\phi, 0) \begin{pmatrix} d_{15}\sin\phi\sin\theta \\ -d_{15}\cos\phi\sin\theta \\ d_{15}(-\sin\phi\cos\phi\cos\theta + \cos\phi\sin\phi\cos\theta) \end{pmatrix}$$

$$= -d_{15}\sin\theta.$$

$$(6.24)$$

Here the sign can be neglected since d_{eff} is squared in (6.10) so it is only its magnitude that is of interest. Thus for this particular case only the d_{15} component of the $\overleftrightarrow{d}$ tensor is important and d_{eff} varies with crystal orientation as a function of θ in the coordinate system shown in Fig. 6.7.

As another example, consider having the same polarization vectors as in the above example but a crystal with T_d symmetry. Using the d tensor for T_d symmetry in Table 6.2 in (6.24) gives the result

$$d_{\text{eff}} = (-\sin\phi, \cos\phi, 0) \begin{pmatrix} 0 & 0 & 0 & d_{14} & 0 & 0 \\ 0 & 0 & 0 & 0 & d_{14} & 0 \\ 0 & 0 & 0 & 0 & 0 & d_{14} \end{pmatrix} \begin{pmatrix} -\sin\phi\cos\phi\cos\theta \\ \cos\phi\sin\phi\cos\theta \\ 0 \\ -\cos\phi\sin\theta \\ \sin\phi\sin\theta \\ (\cos^2\phi - \sin^2\phi)\cos\theta \end{pmatrix}.$$

Thus, using a trig identity,

$$d_{\text{eff}} = d_{14}\sin 2\phi \sin\theta.$$

In this case it is the d_{14} component of the nonlinear optical tensor that is important. In addition, the magnitude of d_{eff} varies not only with the angle θ but also with the azimuthal angle ϕ. Thus the selection of the direction of propagation and polarization is critical in maximizing the intensity of the frequency double light beam. Values of d_{eff} for other combinations of polarization directions are discussed further below.

6.3 Index Matching

The importance of phase matching in maximizing the intensity of the second-harmonic light beam was discussed in Sect. 6.1. Several different types of configurations that result in birefringent phase matching are discussed in this section. The key is finding the right geometry for the directions of travel and directions of polarizations of the incident light beam components and the second-harmonic light beam that ensures that their phase velocities are the same. This requires that $n_1 \approx n_3$. This can be accomplished in uniaxial or biaxial crystals by using their birefringence properties. In these crystals there are two values of the refractive index for each direction of propagation, corresponding to the two allowed orthogonally polarized modes. By orienting the crystal for an appropriate direction of propagation and choosing an appropriate direction of polarization, it is possible to obtain the desired index matching so $\Delta k=0$ in (6.6). Figure 6.8 shows a simple picture of how this can be accomplished by taking advantage of the normal dispersion for a negative crystal ($n_o > n_e$). Note that the value of $n^e(\theta)$ varies between the values of n_o and n_e depending on the angle θ.

As an example, consider the case of uniaxial crystals that have an indicatrix that is an ellipsoid of revolution with the optic axis being the axis of rotation. This was shown previously in Fig. 5.2 and repeated here in Fig. 6.9 with additional information included about the values of the indices of refraction and the two directions of polarization. I_i is the intensity of the incident light beam and $\mathbf{k}_i$ is its direction of propagation. For frequency doubling of one incident light beam, all the propagation directions of interest are co-linear as shown in Fig. 6.2. A plane perpendicular to this direction is shown in the figure. The intersection of this plane with the indicatrix is an ellipse with its two axes parallel to the two directions of polarization and their lengths equal to the value of the refractive index in that direction. Note that the value of the refractive index for polarization perpendicular to the optic axis (ordinary direction) does not change with the direction of propagation, but the

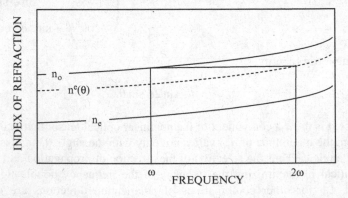

Fig. 6.8 Matching the refractive index of the incident beam at frequency ω and the SHG beam at twice this frequency for a negative crystal

Fig. 6.9 Indicatrix ellipsoid
for a positive crystal

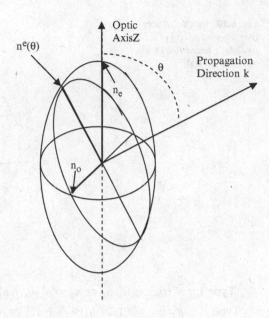

extraordinary direction of propagation has an index of refraction that varies as given
by (6.16) which can be rewritten as

$$n^e(\theta) = \frac{n_o n_e}{\left[(n_o)^2 \sin^2 \theta + (n_e)^2 \cos^2 \theta\right]^{1/2}}. \qquad (6.25)$$

One way to demonstrate phase matching for two wavelengths of light with
different polarizations is shown in Fig. 6.10. For a monochromatic source at the
center, a wavefront for an ordinary ray polarized perpendicular to the optic axis
expands as a sphere but a wavefront for the extraordinary ray expands as an
ellipsoid. The distance of the surface from the origin along the direction of the
wave vector is the magnitude of the refractive index. Along the optic axis the o and
e rays propagate with the same velocity. These index surfaces are shown in
Fig. 6.10 for two wavelengths. If n_3^e refers to the extraordinary ray for the second-harmonic wavelength and n_1^o is for the ordinary ray of the fundamental, index
matching or phase matching is satisfied for propagation at an angle θ_m from the
optic axis of the crystal as shown in Fig. 6.10. Thus angle-tuned SHG depends on
wavelength, propagation direction, and polarization.

There are two ways for achieving phase matching depending on the phase
velocities of the rays. Type I is when the two components of the fundamental
wave have the same direction of polarization the frequency double wave is polarized orthogonal to that direction. Type II is when the two components of the
fundamental wave are orthogonally polarized and the frequency doubled wave
has one or the other directions of polarization. Thus,

Fig. 6.10 Index surfaces for two wavelengths, λ_1 (*solid line*); λ_3 (*dotted line*) for a negative crystal

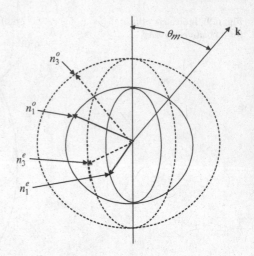

Type I : $\quad n_3^e(\theta_m) = n_1^o, n_1^o$ or $n_3^o = n_1^e(\theta_m), n_1^e(\theta_m)$,

Type II : $\quad n_3^e(\theta_m) = (1/2)\left[n_1^e(\theta_m) + n_1^o\right]$ or $n_3^o = (1/2)\left[n_1^e(\theta_m) + n_1^o\right]$, $\qquad$ (6.26)

where the first choice is for a negative crystal and the second choice is for a positive crystal.

Another convenient way to look at phase matching is to consider an ordinary and an extraordinary wave expanding outward from the same origin. This is shown in Fig. 6.11 for a negative crystal. As an example of the first case of Type I phase matching, two o rays at frequency ω are matched to an e ray at frequency 2ω similar to the situation shown in Figs. 6.8 and 6.10. The index ellipsoids for Type I phase matching in a negative are shown in Fig. 6.11. As an example of the second case of Type I phase matching, an o ray and an e ray at ω are matched to an e ray at 2ω.

Note that the SHG efficiency can depend on the azimuthal angle as well as the phase matching angle θ_m because d_{eff} is expressed as one or several coefficients of the d tensor and the angles defining the directions of wave propagations and polarizations. A typical example of the dependence on both angles is $d_{eff}=d_{14}\sin2\phi\sin\theta$ as derived above.

An expression for the phase matching angle can be found by substituting one of the expressions in (6.26) into (6.25). This gives

$$\sin^2\theta_m = \frac{\left(n_1^o\right)^{-2}-\left(n_3^o\right)^{-2}}{\left(n_{3e}\right)^{-2}-\left(n_3^o\right)^{-2}}. \qquad (6.27)$$

For the two types of Type I phase matching, (6.27) becomes

Fig. 6.11 Type I phase
matching for a negative
crystal

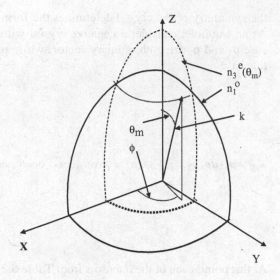

$$n_1^o\ n_1^o\ n_3^e\ :\quad \sin\theta_m = \left(\frac{n_3^e}{n_1^o}\right)^2 \left(\frac{\left(n_3^o\right)^2 - \left(n_1^o\right)^2}{\left(n_3^o\right)^2 - \left(n_3^e\right)^2}\right),\quad (n_e < n_o),$$

$$n_1^e\ n_1^e\ n_3^o\ :\quad \sin\theta_m = \left(\frac{n_1^e}{n_3^o}\right)^2 \left(\frac{\left(n_3^o\right)^2 - \left(n_1^o\right)^2}{\left(n_1^e\right)^2 - \left(n_1^o\right)^2}\right),\quad (n_e > n_o).$$

$$(6.28)$$

These expressions are more complicated for Type II phase matching.

As an example, consider Type I phase matching in a negative crystal with $n_1^o = 1.507$, $n_1^e = 1.468$, $n_3^o = 1.528$, and $n_3^e = 1.482$. Substituting these numbers into the first expression in (6.28) gives $\theta_m = 65°$.

6.4 Maximizing SHG Efficiency

The derivation of (6.10) showed that phase matching and the effective nonlinear optical coefficient were the two critical parameters in obtaining efficient frequency doubling. Each of these was discussed individually in Sects. 6.2 and 6.3 In this section it is shown how the two can be combined to obtain maximum SHG efficiency.

The key to maximizing I_{SHG} is the expression for d_{eff} given in (6.11). This can be written in tensor form using the polarization vector components from (6.23) based on the coordinate system shown in Fig. 6.7. The phase matching conditions listed in (6.26) determine the components of the **p** vectors used in the equation and the symmetry of the crystal determines the form of the d matrix from Table 6.2. As an example, consider a negative crystal with Type I phase matching. In this case $\mathbf{p}_1$ and $\mathbf{p}_2$ are both ordinary vectors while $\mathbf{p}_3$ is an extraordinary vector. For this case

$$d_{eff} = \vec{p}_3 \cdot \vec{\vec{dp}}_1 \vec{p}_2 = \vec{p}_e \cdot \vec{\vec{dp}}_o \vec{p}_o = (\cos\theta_m \cos\phi, \cos\theta_m \sin\phi, -\sin\theta_m) \cdot \vec{\vec{d}} \begin{pmatrix} \sin^2\phi \\ \cos^2\phi \\ 0 \\ 0 \\ 0 \\ -2\sin\phi\cos\phi \end{pmatrix}.$$

$$(6.29)$$

At this point, each of the d tensors from Table 6.2 can be used in this expression to find the effective nonlinear optical coefficient for each symmetry class of uniaxial crystals.

Using the d tensor for crystal symmetry classes C_4, C_6, C_{4v} and C_{6v} in (6.29) gives

$$d_{eff} = (\cos\theta_m \cos\phi, \cos\theta_m \sin\phi, -\sin\theta_m) \cdot \begin{pmatrix} 0 & 0 & 0 & 0 & d_{15} & 0 \\ 0 & 0 & 0 & d_{15} & 0 & 0 \\ d_{15} & d_{15} & d_{33} & 0 & 0 & 0 \end{pmatrix} \begin{pmatrix} \sin^2\phi \\ \cos^2\phi \\ 0 \\ 0 \\ 0 \\ -2\sin\phi\cos\phi \end{pmatrix}$$

$$= (\cos\theta_m \cos\phi, \cos\theta_m \sin\phi, -\sin\theta_m) \cdot \begin{pmatrix} 0 \\ 0 \\ d_{15} \end{pmatrix}$$

$$= -d_{15}\sin\theta_m.$$

Note that this is the same result as was obtained in the previous example of oeo polarizations with the same d tensor. For this case the effective nonlinear optical coefficient depends only on the angle θ_m as determined by (6.28). It would be maximum if $\theta_m = 90°$. The azimuthal direction of propagation of the light beams can be any direction in the XY plane.

Next consider an example of the same phase matching conditions for the polarization vectors given in (6.29) but for crystals with D_{2d} symmetry. In this case (6.29) becomes

$$d_{eff} = (\cos\theta_m \cos\phi, \cos\theta_m \sin\phi, -\sin\theta_m) \cdot \begin{pmatrix} 0 & 0 & 0 & d_{14} & 0 & 0 \\ 0 & 0 & 0 & 0 & d_{14} & 0 \\ 0 & 0 & 0 & 0 & 0 & d_{14} \end{pmatrix} \begin{pmatrix} \sin^2\phi \\ \cos^2\phi \\ 0 \\ 0 \\ 0 \\ -2\sin\phi\cos\phi \end{pmatrix}$$

$$= (\cos\theta_m \cos\phi, \cos\theta_m \sin\phi, -\sin\theta_m) \cdot \begin{pmatrix} 0 \\ 0 \\ -2d_{14}\sin\phi\cos\phi \end{pmatrix}$$

$$= d_{14}\sin\theta_m \sin 2\phi.$$

Once again this is the same result as obtained in a previous example for oeo polarizations with the same d tensor. For this case the phase matching angle is again determined by (6.27) and would be maximized if $\theta_m = 90°$ but in addition the azimuthal angle must be $\phi = 45°$ for d_{eff} to be maximum. Thus the direction of propagation of the light beams is in the XY plane half way between the X and Y axes.

This same procedure can be carried out for all of the symmetry classes for uniaxial crystals and the results are summarized in Table 6.3. Note that only one or two of the components of the d tensor contribute to d_{eff} for a specific crystal symmetry class. Also note that for many symmetry classes, d_{eff} is maximized for $\theta_m = 90°$. This is important for practical applications because diffraction and beam divergence result in a dependence of the propagation vector on θ. If $\theta_m = 90°$ this "walk-off" problem disappears. Thus 90° phase matching is referred to as *noncritical phase matching* and this allows more effective SHG to occur. Thus crystals that can maximize d_{eff} with noncritical phase matching are important in frequency doubling applications.

Table 6.3 Effective nonlinear optical coefficients for uniaxial, negative crystals for Type I phase matching with both fundamental wave components polarized in the ordinary direction and the SHG wave polarized in the extraordinary direction (with Kleinman symmetry)

Symmetry class	d_{eff}
D_4, D_6	0
C_4, C_{4v}, C_6, C_{6v}	$-d_{15}\sin\theta_m$
D_{2d}	$d_{14}\sin\theta_m\sin 2\phi$
C_3	$-(d_{11}\cos3\phi - d_{16}\sin3\phi)\cos\theta_m - d_{15}\sin\theta_m$
D_3	$-d_{11}\cos\theta_m\cos 3\phi$
D_{3h}	$d_{16}\cos\theta_m\sin 3\phi$
C_{3v}	$d_{16}\cos\theta_m\sin 3\phi - d_{15}\sin\theta_m$
C_{3h}	$-(d_{11}\cos3\phi - d_{16}\sin3\phi)\cos\theta_m$
S_4	$(d_{15}\cos2\phi + d_{14}\sin2\phi)\sin\theta_m$

For Type I phase matching in a positive crystal, $\mathbf{p}_1$ and $\mathbf{p}_2$ are both extraordinary vectors while $\mathbf{p}_3$ is an ordinary vector. For this case

$$d_{eff} = \vec{p}_3 \cdot \vec{\vec{dp}}_1 \vec{p}_2 = \vec{p}_o \cdot \vec{\vec{dp}}_e \vec{p}_e = (\cos\theta\cos\phi, \cos\theta\sin\phi, -\sin\theta) \cdot \vec{\vec{d}} \begin{pmatrix} \cos^2\phi\cos^2\theta \\ \sin^2\phi\cos^2\theta \\ \sin^2\theta \\ -\sin2\theta\sin\phi \\ -\sin2\theta\cos\phi \\ \cos^2\theta\sin2\phi \end{pmatrix}.$$

The d tensors from Table 6.2 are substituted into this expression to obtain values for d_{eff} for each uniaxial crystal system. Using the same procedure demonstrated above gives the results in Table 6.4. Note that for this polarization orientation there are fewer opportunities to maximize d_{eff} through 90° phase matching.

When Type II phase matching as defined in (6.26) is used, different combinations for the polarization vectors from (6.23) are needed for (6.29). The same procedure used above can be carried out using these new polarization vectors. In this case the two components of the fundamental wave have orthogonal ordinary and extraordinary polarizations. If the frequency doubled wave is polarized in the ordinary direction, the values of d_{eff} for each crystal system are the same as those shown in Table 6.3. If the frequency doubled wave is polarized in the extraordinary direction the values of d_{eff} are the same as those given in Table 6.4. For example, for a positive crystal having D_{2d} symmetry, the d_{eff} can be maximized through either Type I or Type II phase matching using the expressions

$$d_{14}\sin2\theta_m\cos2\phi, \quad \text{Type I,}$$
$$d_{14}\sin\theta_m\sin2\phi, \quad \text{Type II.}$$

For both cases the phase matching angle is determined by (6.27). It might be possible to achieve noncritical phase matching using a Type II configuration but

Table 6.4 Effective nonlinear optical coefficients for uniaxial, positive crystals for Type I phase matching with both fundamental wave components polarized in the extraordinary direction and the SHG wave polarized in the ordinary direction (with Kleinman symmetry)

Symmetry class	d_{eff}
D_4, D_6	0
C_4, C_{4v}, C_6, C_{6v}	0
D_{2d}	$d_{14}\sin2\theta_m\cos2\phi$
C_3	$-(d_{11}\sin3\phi+d_{16}\cos3\phi)\cos^2\theta_m$
D_3	$-d_{11}\cos^2\theta_m\sin3\phi$
D_{3h}	$-d_{16}\cos^2\theta_m\cos3\phi$
C_{3v}	$-d_{16}\cos^2\theta_m\cos3\phi$
C_{3h}	$-(d_{11}\sin3\phi+d_{16}\cos3\phi)\cos^2\theta_m$
S_4	$-(d_{15}\sin2\phi+d_{14}\cos2\phi)\sin2\theta_m$

not Type I. For Type I phase matching the propagation direction should have $\phi=0°$ or 90°. For Type II phase matching the propagation direction should have $\phi=45°$ or 135°. This demonstrates the importance of the azimuthal angle for different types of phase matching. The same procedures can be used for biaxial crystal systems but the resulting expressions depend critically on the choice crystallographic axes directions compared to laboratory coordinates and propagation directions [2].

In summary, maximizing the efficiency of second-harmonic generation as expressed in (6.14) requires maximizing d_{eff}. The parameters involved in doing this are the choice of the nonlinear material, the wavelength of the incident light beam, the direction of propagation, and the polarization directions of the initial and final light beams. These are interdependent and in some cases not subject to arbitrary choice. For example, a laser of a specific wavelength may be required or only a specific nonlinear crystal is available. Given these fixed parameters, different types of polarization combinations and propagation directions can be chosen. The resulting expressions for d_{eff} in Tables 6.3 and 6.4 can be optimized using the appropriate values for θ_m and ϕ. Note an estimate for θ_m that will maximize phase matching can be found from (6.28). This may not be the value of θ that will maximize the trigonometric expression for d_{eff}. It is generally necessary to tune the temperature and θ to obtain the maximum conversion efficiency.

Finally, examples of measured values of nonlinear optical coefficients are given for some common frequency doubling crystals in Table 6.5. These measured values are given in SI units of (pm/V). To convert to cgs units of $(cm^3/erg)^{1/2}$ the values in the table should be multiplied by 2.39×10^{-9}. Different measurements have yielded slightly different values for the nonlinear optical coefficients depending on the crystal quality and experimental conditions. Thus there are some discrepancies in the literature, but the results are all close to those listed in the table. The first row of each entry lists the primary phase matching configuration and relevant nonlinear coefficients while the second row lists other phase matching configurations that have been reported and values of other coefficients that have been measured. Tables 6.3 and 6.4 can be used with these values to determine d_{eff}. Equation (6.28) can be used with the values of the refractive indices to determine the phase matching angle. However, the angle listed is experimentally measured and the theoretically predicted angle is only approximate due to inaccuracies in the measurements of the indices of refraction.

6.5 Two-Photon Absorption

In optical spectroscopy, the simultaneous absorption of two photons is made possible by the use of high power laser sources. This can occur using a single laser source producing indistinguishable photons or by using two different lasers that produce two different photons. Experimental two-photon absorption spectroscopy using broadband tunable lasers has become a useful tool for studying the properties of the excited states of atomic, molecular, and solid state systems. A two-

Table 6.5 Selected values of nonlinear optical coefficients for some common crystals used in frequency doubling. Measurements were made at the Nd:YAG laser wavelength of 1,064nm [1,2,4,5]

Crystal (symmetry)	Indices of refraction	Phase matching	Nonlinear optical coefficients (pm/V)
Lithium niobate ($n_o > n_e$)	$n_1^o = 2.2340, n_3^o = 2.3251$	$\theta_{ooe} = 90°$	$d_{15} = -5.95, d_{16} = 2.8$
LiNbO₃:MgO (5%) (C_{3v})	$n_1^e = 2.1554, n_3^e = 2.2330$	eoe, oee	$d_{33} = -34.4$
KTP ($n_{2\omega} > n_\omega$)	$n_{1x} = 1.738, n_{3x} = 1.778$	$\theta_{eoe} = 30°$	$d_{15} = 6.1, d_{24} = 7.6$
KTiOPO₄ (C_{2v})	$n_{1y} = 1.746, n_{3y} = 1.789$	oeo, ooe	$d_{33} = 13.7$
	$n_{1z} = 1.830, n_{3z} = 1.889$		
ADP ($n_o > n_e$)	$n_1^o = 1.507, n_3^o = 1.528$	$\theta_{oee} = 62°$	$d_{14} = 0.53$
NH₄H₂PO (D_{2d})	$n_1^e = 1.468, n_3^e = 1.482$	ooe, eoe	
LBO ($n_{2\omega} > n_\omega$)	$n_{1x} = 1.566, n_{3x} = 1.579$	$\theta_{ooe} = 90°$	$d_{24} = 1.2$
LiB₃O₅ (C_{2v})	$n_{1y} = 1.591, n_{3y} = 1.607$	oeo, eoo,	$d_{15} = 1.1, d_{33} = 0.07$
	$n_{1z} = 1.606, n_{3z} = 1.621$	eoe, eeo	
Quartz ($n_e > n_0$)	$n_1^o = 1.534, n_3^o = 1.5468$		$d_{11} = 0.37$
SiO₂ (D_3)	$n_1^e = 1.543, n_3^e = 1.5560$		
KNbO₃ ($n_{2\omega} > n_\omega$)	$n_{1x} = 2.114, n_{3x} = 2.199$	$\theta_{ooe} = 90°$	$d_{24} = -13.2$
	$n_{1y} = 2.220, n_{3y} = 2.319$		
	$n_{1z} = 2.258, n_{3z} = 2.377$		
(C_{2v})		eeo, eoo	$d_{15} = 11.5, d_{33} = -20$
LiIO₃ ($n_o > n_e$)	$n_1^o = 1.857, n_3^o = 1.898$	$\theta_{ooe} = 30°$	$d_{15} = -7.1$
(C_6)	$n_1^e = 1.717, n_3^e = 1.748$		$d_{33} = -7.0$
BBO ($n_o > n_e$)	$n_1^o = 1.655, n_3^o = 1.675$	$\theta_{ooe} = 23°$	$d_{15} = 0.12, d_{16} = 1.8$
β−BaB₂O₄ (C_{3v})	$n_1^e = 1.543, n_3^e = 1.556$	eoe, oee	
KD*P ($n_o > n_e$)	$n_1^o = 1.4928, n_3^o = 1.5085$	$\theta_{eoe} = 54°$	$d_{14} = 0.53$
KD₂PO₄ (D_{2d})	$n_1^e = 1.4555, n_3^e = 1.4690$	oee, ooe	

photon transition involving a real intermediate state is generally referred to as excited state absorption while a true two-photon absorption transition involves a virtual intermediate state. The selection rules for this type of process are different from single-photon transition processes so a two-photon absorption spectrum can provide information about the symmetry of the excited state as discussed below. This section describes the application of two-photon spectroscopy to a molecular complex as an example of this phenomenon.

The spatial positions of the atoms that constitute a particular molecule define its symmetry. The symmetry operations that leave the arrangement of the atoms of the molecule invariant form the symmetry group of the molecule which is given by one of the point group symmetries discussed in Chap. 2. The stationary electronic states of a molecule are described by wavefunctions with specific symmetries. Typically the ground electronic state has the symmetry of the molecular point group while higher lying electronic states have lower symmetries.

As discussed in previous chapters, for one-photon transitions, allowed transitions occur only between states of opposite parity. On the other hand, for two-photon transitions, the odd parity electric dipole moment operator acts twice so the allowed transitions are between the states of the same parity. Therefore the two-photon absorption spectra show lines not seen in the one-phonon spectra. The polarization direction of the excitation beams is critical in determining the allowed transitions as described below and this helps determine the symmetry properties of the excited states of the transitions.

Optical transitions are described through the interaction of an electromagnetic radiation field with the electronic states of the system. As discussed in Sect. 6.1, the electric dipole term in the expansion of the radiation field is the dominant operator inducing the transition. Quantum mechanical perturbation theory is used to show that the probability, or strength, of the process is determined by the matrix element of the transition. For a one-photon process this is expressed by (4.23). For a two-photon process second-order perturbation theory is used and the expression for the transition matrix element becomes [6, 7]

$$M_{ed}^{2p} = \sum_{a,b} \sum_{k} \left[\frac{\langle i|\vec{r}_a|k\rangle \langle k|\vec{r}_b|f\rangle}{\omega_{ki} - \omega_1} + \frac{\langle i|\vec{r}_b|k\rangle \langle k|\vec{r}_a|f\rangle}{\omega_{ki} - \omega_2} \right]. \qquad (6.30)$$

The transition described by this expression is illustrated in Fig. 6.12. It involves photons of two different frequencies ω_1 and ω_2 with the two terms in brackets indicating the contributions depending on which photon was absorbed first.

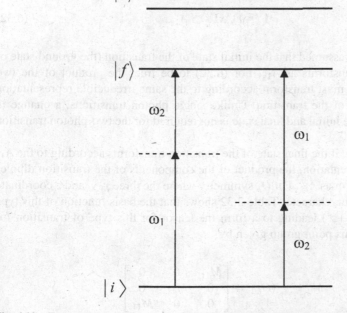

Fig. 6.12 Energy level diagram for two-photon absorption transitions

These contributions differ because different photon energies lead to different values for resonant denominators. The some over k includes all higher lying real states that play the role of virtual intermediate states (dashed lines) in the transition process. The initial and final states are designated by i and f while $\vec{r}_a$ and $\vec{r}_b$ are the transition dipoles for the molecule.

Because of the variation of the transition matrix element in (6.30) with the polarization directions of the two photons, it can best be expressed as a second-rank tensor of the form

$$M_{ed}^{2p} = \begin{bmatrix} M_{11} & M_{12} & M_{13} \\ M_{21} & M_{22} & M_{23} \\ M_{31} & M_{32} & M_{33} \end{bmatrix}, \tag{6.31}$$

where each component represents the contribution to the transition strength from the directions of the transition dipoles in the molecule. The nine elements M_{ij} characterize the symmetry of the excited state of a molecule with specific point group symmetry.

The individual value a specific tensor element indicates its ability to couple the two transition dipoles $\vec{r}_a$ and $\vec{r}_b$. The form of the two-photon absorption tensor depends on the symmetry group of the molecule and the irreducible representations designating the states involved in the transition. Generalizing the one-photon transition expressions in (2.24) and (4.33) to the two-photon absorption case, for the transition matrix in (6.30) to be nonzero

$$\Gamma(\vec{r}_a\vec{r}_b)x\Gamma_f \supset A_{1g} \tag{6.32}$$

where it has been assumed that the initial state of the transition (the ground state of the molecule) transforms as A_{1g}. For (6.32) to be true, the product of the two transition dipoles must transform according to the same irreducible representation of the final state of the transition. Unlike single photon transitions, a change in parity between the initial and final state is not required for the two-photon transition to be allowed.

As an example, if the final state of the transition transforms according to the A_{1g} irreducible representation, the product of the components of the transition dipoles must also transform as A_{1g}. For O_h symmetry where the three x, y, and z coordinate axes are equivalent, character Table 2.32 shows that the basis function of this type for A_{1g} is $(x^2+y^2+z^2)$ leading to a form the tensor for this type of transition for molecules with this point group given by

$$M_{A_{1g}\rightarrow A_{1g}}^{2p}(O_h) = \begin{bmatrix} M_{11} & 0 & 0 \\ 0 & M_{11} & 0 \\ 0 & 0 & M_{11} \end{bmatrix},$$

where $M_{11}=(M_{xx}+M_{yy}+M_{zz})/3$. The same transition in molecules whose symmetry group has only two equivalent axes such as D_{4h} will have a transition tensor of the form

$$M^{2p}_{A_{1g}\rightarrow A_{1g}}(D_{4h}) = \begin{bmatrix} M_{11} & 0 & 0 \\ 0 & M_{11} & 0 \\ 0 & 0 & M_{33} \end{bmatrix}.$$

Here $M_{11}=(M_{xx}+M_{yy})/2$ and $M_{33}=M_{zz}$. (See Tables 2.14 and 3.4.) For symmetries with no equivalent axes the tensor has the form

$$M^{2p}_{A_{1g}\rightarrow A_{1g}}(D_{2h}) = \begin{bmatrix} M_{xx} & 0 & 0 \\ 0 & M_{yy} & 0 \\ 0 & 0 & M_{zz} \end{bmatrix}.$$

If the final state has a wavefunction that transforms according to some irreducible representation other than A_{1g} the off-diagonal elements of the transition tensor that represent orthogonal transition dipoles can be nonzero. For example, the transition to a T_{2g} state in a molecule having O_h symmetry is allowed for the two transition dipoles orient as xy, yz, and xy (see Table 2.32). Thus the transition tensor has the form

$$M^{2p}_{A_{1g}\rightarrow T_{2g}}(O_h) = \begin{bmatrix} 0 & M_{12} & M_{13} \\ M_{12} & 0 & M_{23} \\ M_{13} & M_{23} & 0 \end{bmatrix},$$

where $M_{12}=(M_{xy}+M_{yx})/2$, $M_{13}=(M_{xz}+M_{zx})/2$, and $M_{23}=(M_{yz}+M_{zy})/2$.

The examples given above show the form of the two-photon absorption tensor to be either diagonal or symmetric. This is always the case when the two photons involved are identical. For distinguishable photons, some of the high symmetry groups can have transition tensors that are antisymmetric. This can occur when the wavefunction of the excited state changes sign under the interchange of the x and y axes [8].

The transition strength observed experimentally involves the product of the two-photon absorption tensor with the two photon polarization vectors $\vec{p}_1$ and $\vec{p}_2$

$$I^{2p} = \vec{p}_1 M^{2p}_{ed}\vec{p}_2$$

$$= (p_{1x}, p_{1y}, p_{1z}) \begin{bmatrix} M_{xx} & M_{xy} & M_{xz} \\ M_{yx} & M_{yy} & M_{yz} \\ M_{zx} & M_{zy} & M_{zz} \end{bmatrix} \begin{pmatrix} p_{2x} \\ p_{2y} \\ p_{2z} \end{pmatrix}. \tag{6.33}$$

This form of the expression assumes that the coordinates of all the molecules involved in the sample being studied are in the same fixed alignment with the laboratory coordinate system. This is only true for special cases such as molecular solids or other systems that can be physically aligned. When this is the case, polarized two-photon absorption spectroscopy can determine the form of the transition tensor and thus the symmetry properties of the excited state of the transition. For example, in the case described above for a molecule with O_h symmetry, polarizing both photons along the x direction will provide a contribution to the spectral intensity involving only the M_{xx} tensor component. Doing similar experiments with both photons polarized along the y-axis or the z-axis will show the contributions to the spectral intensity involving the M_{yy} and M_{zz} tensor components. If these three tensor components are found experimentally be nonzero and approximately the same magnitude, it shows that the form of the two-photon absorption tensor is diagonal which means that the excited state of the transition transforms according to the A_{1g} irreducible representation of the O_h symmetry group. On the other hand, if the spectral intensity is essentially zero for these photon polarization conditions other experiments are needed. By polarizing one of the photons in the x direction and the other in the z direction the contribution of the M_{xz} component can be observed. Similarly, polarizing one photon in the x direction and the other in the y direction gives the contribution of the M_{xy} tensor component while polarizing the photons in the y and the z directions measures the contribution of the M_{yz} component. If these three off-diagonal elements are nonzero, the form of the tensor is consistent with an excited state transforming according to the T_{2g} irreducible representation of the O_h point group. In this way a set of two-photon absorption spectra for different combinations of photon polarization directions can be used to determine the irreducible representation according to which the excited state transforms. It should be noted that this procedure is very similar to Raman scattering spectroscopy discussed in Sect. 7.4 and summarized in Table 7.8.

For the situation in which the coordinate axes of the molecule are not aligned with the laboratory coordinates, (6.33) must be modified to include a factor of $\cos\theta_a\cos\theta_b$ where the θ_i are the angles between the laboratory coordinates and the molecular axes. This provides the required transformation between molecular axes and laboratory coordinates. In general, an ensemble of molecules being studied will have random orientations and the expression for transition strength must be averaged over all directions. This limits the amount of detailed information that can be obtained through polarized two-photon absorption spectroscopy, but it is still possible to determine if the transition tensor is diagonal, nondiagonal symmetric, or nondiagonal antisymmetric.

6.6   Problems

1. What is the coherence length for second-harmonic generation at a wavelength of 532 nm if the relevant refractive indices are $n_1=1.738$ and $n_3=1.789$?
2. Derive the form of the nonlinear optical tensor for a crystal with C_{2v} symmetry with and without Kleinman Symmetry.

3. Derive the expression for the effective nonlinear optical coefficient for two incident light waves with ordinary polarization directions and a frequency doubled wave with polarization in the extraordinary direction in a crystal with C_{3v} symmetry.
4. Calculate the phase matching angle with the light polarization conditions given in problem 3 with $n_1^o = 1.738, n_3^o = 1.789$, and $n_3^e = 1.668$.
5. For a crystal with D_{2d} symmetry and noncritical phase matching, what azimuthal angles maximize the effective nonlinear optical coefficient?

References

1. R.L. Sutherland, *Handbook of Nonlinear Optics* (Dekker, New York, 1996)
2. V.G. Dmitriev, G.G. Gurzadyan, D.N. Nikogosyan, *Handbook of Nonlinear Optical Crystals* (Springer, Berlin, 1991)
3. R.W. Munn, C.N. Ironside, *Nonlinear Optical Materials* (Blackie, Glasgow, 1993)
4. R. Guenther, *Modern Optics* (Wiley, New York, 1990)
5. A. Yariv, *Quantum Electronics* (Wiley, New York, 1989)
6. B. Boulanger, J. Zyss, in *International Tables for Crystallography Volume D, Physical Properties of Crystals*, ed. A. Authier (Kluwer, Dordrecht, 2003), p. 178
7. M. Goeppert-Mayer, Ann. Phys. 9, 273 (1931)
8. M.J. Wirth, A. Koskelo, M.J. Sanders, App. Spectros 35, 14 (1981)

Chapter 7
Symmetry and Lattice Vibrations

The other chapters in this book deal with the atoms in a crystalline solid being in their static equilibrium positions. This chapter focuses on the thermal vibrations of the atoms about their equilibrium positions. This motion is treated in the harmonic approximation. The symmetry of the lattice plays an important role in determining how the atoms move. The positions of neighboring atoms can inhibit motion in some directions while facilitating motion in other directions. This results in certain "normal modes" of vibration being allowed and other vibrational modes not allowed. Any state of vibration of the lattice can be expressed as a superposition of normal modes. The energy of the vibrational modes is quantized and can be described by eigenvectors and eigenvalues (frequencies). Each of these modes exhibit specific symmetry and can be associated with one of the irreducible representations of the crystallographic point group.

The quantized vibrational modes are called *phonons*. A phonon is an elementary excitation that can be treated as a quasiparticle in a solid. There are two types of phonons. One involves only a central ion and its nearest ligands. These are called *local mode phonons*. Since these involve atoms in one unit cell, translational symmetry is not important and the point group of the local complex of atoms is used to describe the symmetry of the vibrations. The second type of vibration involves the motion of atoms throughout many unit cells of the lattice. In this case, translational symmetry is important and the vibrations are called *lattice phonons*. The space group of the crystal must be used in analyzing the lattice vibrations. Some places in the literature reserve the term phonon only for extended lattice vibrations and not local modes. Here it is used for all vibrational modes.

Transitions between the quantized energy levels of lattice vibrations can occur through the absorption or emission of electromagnetic radiation. This can be described as the creation or annihilation of specific phonons. The selection rules governing whether specific transitions are allowed or forbidden can be determined through group theory considerations.

Sections 7.1 and 7.2 treat local modes and lattice phonons separately and demonstrate the importance of symmetry and group theory in analyzing the vibrational properties of solids. Section 7.3 summarizes the quantum mechanical treatment of transitions among vibrational energy levels and the selection rules determined by symmetry. Section 7.4 describes the inelastic scattering of light in

R.C. Powell, *Symmetry, Group Theory, and the Physical Properties of Crystals*,
Lecture Notes in Physics 824, DOI 10.1007/978-1-4419-7598-0_7,
© Springer Science+Business Media, LLC 2010

which phonons are created or destroyed. This *Raman scattering* is characterized as a second-rank tensor and the tensor mathematics described in Chap. 3 is used to determine the selection rules for specific phonon modes involved with this type of process.

7.1 Symmetry and Local Mode Vibrations

Some vibrational modes are localized at specific points in the crystal lattice and can be described by the motion atoms around these points. This can occur if the physical system involves a point defect as discussed in Chap. 4 or if there is strong bonding among a local complex of atoms. In this case, phonon momentum and long range translational symmetry can be ignored and the point symmetry group of the local site can be used.

The best way to understand the role of symmetry in lattice vibrations is through specific examples. Consider seven ions arranged in an octahedral configuration as shown in Fig. 7.1. Cartesian coordinates can be attached to each ion and three sets of these are shown in part a of the figure as well as the x,y,z coordinate directions for the system. The atoms can be designated by their positions with the central atom being 0, the atoms along the $+$ and $-$ z axes being 1 and 6, respectively, the atoms along the $+$ and $-$ x axes being 2 and 4, respectively, and the atoms along the $+$ and $-$ y axes being 3 and 5, respectively. The character table for the O_h symmetry group is repeated in Table 7.1 for convenience. The double group operations and half-integer representations have been omitted for simplicity.

The motion of the atoms of this complex transforms as a reducible representation Γ_M of the O_h symmetry group with the coordinate vectors attached to each atom forming the basis vectors of this representation. In order to find the characters of this representation the transformation matrices for each symmetry operation of the group must be determined. The trace of each matrix is the character of Γ_M for that

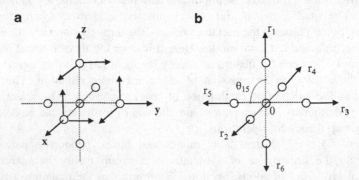

Fig. 7.1 Complex of seven atoms in an octahedral configuration. (a) Laboratory coordinate directions; (b) internal coordinate directions

Table 7.1 Character table for point group O_h

O_h	E	$8C_3$	$6C_2$	$6C_4$	$3C_4^2$	i	$6S_4$	$8S_6$	$3\sigma_h$	$6\sigma_d$	Basis components
A_{1g}	1	1	1	1	1	1	1	1	1	1	$x^2+y^2+z^2$
A_{2g}	1	1	-1	-1	1	1	-1	1	1	-1	
E_g	2	-1	0	0	2	2	0	-1	2	0	$(2z^2-x^2-y^2, x^2-y^2)$
T_{1g}	3	0	-1	1	-1	3	1	0	-1	-1	(R_x, R_y, R_z)
T_{2g}	3	0	1	-1	-1	3	-1	0	-1	1	(xz, yz, xy)
A_{1u}	1	1	1	1	1	-1	-1	-1	-1	-1	
A_{2u}	1	1	-1	-1	1	-1	1	-1	-1	1	
E_u	2	-1	0	0	2	-2	0	1	-2	0	
T_{1u}	3	0	-1	1	-1	-3	-1	0	1	1	(x, y, z)
T_{2u}	3	0	1	-1	-1	-3	1	0	1	-1	
Γ_M	21	0	-1	3	-3	-3	-1	0	5	3	$A_{1g}+E_g+T_{1g}+T_{2g}+3T_{1u}+T_{2u}$
Γ_T	3	0	-1	1	-1	-3	-1	0	1	1	T_{1u}
Γ_R	3	0	-1	1	-1	3	1	0	-1	-1	T_{1g}
Γ_V	15	0	1	1	-1	-3	-1	0	5	3	$A_{1g}+E_g+T_{2g}+2T_{1u}+T_{2u}$
Γ_r	6	0	0	2	2	0	0	0	4	2	$A_{1g}+E_g+T_{1u}$
Γ_θ	12	0	2	0	0	0	0	0	4	2	$A_{1g}+E_g+T_{2g}+T_{1u}+T_{2u}$

operation. Since there are $3N$ coordinate vectors the transformation matrix has the dimensions $3N \times 3N$ with N the number of ions in the complex. Since $N=7$ for the octahedral complex shown in Fig. 7.1, the transformations are represented by 21×21 matrices of the form

$$
\begin{array}{c}
\\ x_0 \\ y_0 \\ z_0 \\ x_1 \\ y_1 \\ z_1 \\ \vdots \\ x_6 \\ y_6 \\ z_6
\end{array}
\begin{pmatrix}
\begin{array}{ccccccccc}
x_0 & y_0 & z_0 & x_1 & y_1 & z_1 & \cdots & x_6 & y_6 & z_6 \\
\cos\alpha & -\sin\alpha & 0 & - & - & - & \cdots & - & - & - \\
\sin\alpha & \cos\alpha & 0 & - & - & - & \cdots & - & - & - \\
0 & 0 & \pm 1 & - & - & - & \cdots & - & - & - \\
- & - & - & \cos\alpha & -\sin\alpha & 0 & \cdots & - & - & - \\
- & - & - & \sin\alpha & \cos\alpha & 0 & \cdots & - & - & - \\
- & - & - & 0 & 0 & \pm 1 & \cdots & - & - & - \\
\vdots & \vdots & \vdots & \vdots & \vdots & \vdots & & \vdots & \vdots & \vdots \\
- & - & - & - & - & - & \cdots & \cos\alpha & -\sin\alpha & 0 \\
- & - & - & - & - & - & \cdots & \sin\alpha & \cos\alpha & 0 \\
- & - & - & - & - & - & \cdots & 0 & 0 & \pm 1
\end{array}
\end{pmatrix}
$$

where the symmetry transformation of the coordinate vectors of each individual atom is given by the 3×3 matrix that was discussed in (2.5). The $+1$ or -1 is for proper and improper rotations, respectively. The submatrices that appear along the diagonal are for atoms that do not change position under the symmetry operation while the off-diagonal submatrices are for ions that change from one position to another in the complex. The trace of this matrix is

$$\chi_M(\alpha) = N_u(\pm 1 + 2\cos\alpha), \tag{7.1}$$

where N_u represents the number of atoms that remain unchanged under the symmetry operation.

Equation (7.1) can be applied to the symmetry operations of the O_h point group given in Table 7.1. For the E operation all seven atoms remain in their positions, it is a proper rotation, and cos $\alpha = 1$ so $\chi_M(E) = 21$. For the C_3 rotation of 120° about the body diagonal of the cube only the atom at position 0 remains unchanged. It is a proper rotation and cos $120° = -1/2$ so $\chi_M(C_3) = 0$. The six C_2 operations that go through the center of one edge of the cube diagonally through the center and opposite edge center leave only the central atom unchanged. It is a proper rotation and cos $180° = -1$. Thus the character is $\chi_M(C_2) = -1$. For the six C_4 operations the three atoms along the rotation axis remain unchanged. It is a proper rotation and cos $90° = 0$. Equation (7.1) gives $\chi_M(C_4) = 3$. The six S_4 operations are improper and leave only the central atom unchanged so $\chi_M(S_4) = -1$. The results of these calculations for all of the symmetry elements of the O_h point group are given in Table 7.1 as the characters of the Γ_M representation.

The motion representation Γ_M is a reducible representation for the O_h point group. Table 7.1 shows how this is reduced in terms of eight irreducible representations of the group. This reduction can be accomplished either by inspection or by using (2.10). The motion representation includes all degrees of freedom, translation, rotation, and vibration. Since the interest here is in the vibrational degrees of freedom the translation and rotation modes must be eliminated. The translation motion of the entire complex transforms as a vector and in O_h symmetry this forms the basis of the T_{1u} irreducible representation. The rotation mode for the entire system transforms as the rotational axes which form the basis for the T_{1g} irreducible representation. Subtracting these two irreducible representations from the total motional representation leaves the vibrational representation Γ_V. As shown in Table 7.1 there are 15 normal modes of vibration for this octahedral complex designated by six irreducible representations of the O_h symmetry group, one nondegenerate mode, one doubly degenerate mode, and four triply degenerate modes.

In order to determine the directions of motion of each of the atoms for each of these normal modes of vibration, it is useful to describe their positions in a different set of coordinates known as *symmetry coordinates*. These internal coordinates for the octahedral complex are shown in Fig. 7.2b. For this complex there are 18 symmetry coordinates. Six of them are radial vectors from the central atom to each of the surrounding atoms labeled r_1 through r_6. The other 12 are angles between specific atoms labeled θ_{12}, θ_{13}, θ_{14}, θ_{15}, θ_{23}, θ_{34}, θ_{45}, θ_{52}, θ_{62}, θ_{63}, θ_{64}, and θ_{65}. Only one of these angles is shown as an example in Fig. 7.2b. Next the reducible representations for the r vectors Γ_r and the angles Γ_θ must be found. The symmetry transformation matrix for the r coordinates is a 6×6 matrix with columns ranging from r_1 to r_6 before the transformation and the rows ranging from r_1 to r_6 after the operation. A value of 1 appears in the ij matrix element when r_i transforms to r_j and the other elements are 0. The matrix elements along the diagonal are 1 when the symmetry operation does not change that r coordinate and 0 if the coordinate is changed. Thus the trace of the matrix equals the number

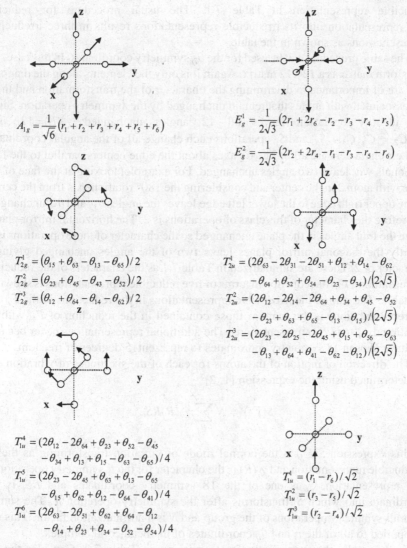

$$A_{1g} = \frac{1}{\sqrt{6}}(r_1 + r_2 + r_3 + r_4 + r_5 + r_6)$$

$$E_g^1 = \frac{1}{2\sqrt{3}}(2r_1 + 2r_6 - r_2 - r_3 - r_4 - r_5)$$

$$E_g^2 = \frac{1}{2\sqrt{3}}(2r_2 + 2r_4 - r_1 - r_3 - r_5 - r_6)$$

$$T_{2g}^1 = (\theta_{15} + \theta_{63} - \theta_{13} - \theta_{65})/2$$

$$T_{2g}^2 = (\theta_{23} + \theta_{45} - \theta_{43} - \theta_{52})/2$$

$$T_{2g}^3 = (\theta_{12} + \theta_{64} - \theta_{14} - \theta_{62})/2$$

$$T_{2u}^1 = (2\theta_{63} - 2\theta_{31} - 2\theta_{51} + \theta_{12} + \theta_{14} - \theta_{62}$$
$$-\theta_{64} + \theta_{52} + \theta_{54} - \theta_{23} - \theta_{34})/(2\sqrt{5})$$

$$T_{2u}^2 = (2\theta_{12} - 2\theta_{14} - 2\theta_{64} + \theta_{34} + \theta_{45} - \theta_{52}$$
$$-\theta_{23} + \theta_{63} + \theta_{65} - \theta_{13} - \theta_{15})/(2\sqrt{5})$$

$$T_{2u}^3 = (2\theta_{23} - 2\theta_{25} - 2\theta_{45} + \theta_{15} + \theta_{56} - \theta_{63}$$
$$-\theta_{13} + \theta_{64} + \theta_{41} - \theta_{62} - \theta_{12})/(2\sqrt{5})$$

$$T_{1u}^4 = (2\theta_{12} - 2\theta_{64} + \theta_{23} + \theta_{52} - \theta_{45}$$
$$-\theta_{34} + \theta_{13} + \theta_{15} - \theta_{63} - \theta_{65})/4$$

$$T_{1u}^5 = (2\theta_{23} - 2\theta_{45} + \theta_{63} + \theta_{13} - \theta_{65}$$
$$-\theta_{15} + \theta_{62} + \theta_{12} - \theta_{64} - \theta_{41})/4$$

$$T_{1u}^6 = (2\theta_{63} - 2\theta_{51} + \theta_{62} + \theta_{64} - \theta_{12}$$
$$-\theta_{14} + \theta_{23} + \theta_{34} - \theta_{52} - \theta_{54})/4$$

$$T_{1u}^1 = (r_1 - r_6)/\sqrt{2}$$

$$T_{1u}^2 = (r_3 - r_5)/\sqrt{2}$$

$$T_{1u}^3 = (r_2 - r_4)/\sqrt{2}$$

Fig. 7.2 Normal modes of vibration of an O_h complex of atoms

of r coordinates left unchanged by the symmetry operation. Since there are six diagonal elements $\chi_r(E)=6$. Each of the C_4 operations leaves the two r coordinates pointed along the fourfold rotation axis invariant while changing the other four. Thus $\chi_r(C_4)=2$. The same is true for the $C_2 = C_4^2$ operations. The C_2 operations around the edge center to edge center diagonal axes change all the r coordinates as do the C_3, I, S_4, and S_6 operations. The horizontal mirror planes leave the four r coordinates in the plane invariant while the diagonal mirror planes leave the two r coordinates in the plane invariant. These characters appear as the Γ_r

reducible representation in Table 7.1. The usual procedure for reducing this representation into its irreducible representations results in three irreducible representations as shown in the table.

The same procedure can be used for the θ_{ij} symmetry coordinates. In this case, the transform matrix is a 12×12 matrix. Again it is only the elements along the diagonal that are of importance in determining the character of the transformation and these are associated with angles that remain unchanged by the symmetry operation. Since the identity operation leaves all 12 angles unchanged $\chi_\theta(E) = 12$. The $C_4, C_2 = C_4^2, C_3, i, S_4$, and S_6 operations each change all of the angular coordinates so their characters are all 0. The C_2 axes about the edge centers parallel to the face diagonals will leave two angles unchanged. For example, looking at the face of the cube with atom 2 in the center and considering the $180°$ rotation axis from the center of the upper right edge to the lower left edge leaves the angles θ_{13} and θ_{56} unchanged. Therefore the character of this class of operations is 2. The horizontal mirror planes leave the four angles in the plane unchanged so the character of these operations is 4. Finally the diagonal mirror planes leave two of the angles unchanged giving a character of 2. These are summarized in Table 7.1 as the characters of the reducible representation Γ_θ. Its reduction in terms of five reducible representations is shown in the table. The sum of the irreducible representations in the reductions of the Γ_r and Γ_θ representations is the same as those contained in the reduction of Γ_V with an additional A_{1g} and E_g representations. The additional representations occur because we have chosen 18 symmetry coordinates to represent 15 degrees of freedom.

The direction of motion of the atoms for each of the six modes of vibration can be determined using the expression [1, 2]

$$S(\Gamma_\gamma) = N \sum_R \chi_\gamma(R) R S_i. \tag{7.2}$$

In this expression $S(\Gamma_\gamma)$ is the normal mode of vibration transforming as the Γ_γ irreducible representation and $\chi_\gamma(R)$ is the character for the R symmetry operation of this representation. S_i is one of the 18 symmetry coordinates and $R S_i$ is the coordinate into which S_i transforms after the symmetry operation R. The sum is over all symmetry operations of the group and N is a normalization factor. This can be applied to all of the r_i and θ_{ij} coordinates of the octahedral complex.

For example, the way r_1 transforms is given in Table 7.2. Consider the six rotations of $+90°$ and $-90°$ about the x, y, and z axes. From Fig. 7.1b it is clear that the fourfold rotation about x transforms r_1 into r_3 or r_5 and a similar rotation about the y axis transforms r_1 into r_2 or r_4. The two C_4 rotations about the z-axis leave r_1 invariant so a 2 appears in this element of the table. None of the six C_4 operations transform r_1 into r_6 so a 0 appears in this element of the table. Using this information in (7.2) along with the characters of the A_{1g} vibrational mode gives

$$S(A_{1g}) = N \sum_R \chi_{A_{1g}}(R) R r_1 = N8(r_1 + r_2 + r_3 + r_4 + r_5 + r_6).$$

Table 7.2 Transformation of the r_1 coordinate in O_h symmetry

R	Rr_1					
	r_1	r_2	r_3	r_4	r_5	r_6
E	1	0	0	0	0	0
$8C_3$	0	2	2	2	2	0
$6C_2$	0	1	1	1	1	2
$6C_4$	2	1	1	1	1	0
$3C_4^2$	1	0	0	0	0	2
i	0	0	0	0	0	1
$6S_4$	0	1	1	1	1	2
$8S_6$	0	2	2	2	2	0
$3\sigma_h$	2	0	0	0	0	1
$6\sigma_d$	2	1	1	1	1	0

This can be normalized to give

$$S(A_{1g}) = \frac{1}{\sqrt{6}}(r_1 + r_2 + r_3 + r_4 + r_5 + r_6). \tag{7.3}$$

This describes the vibrational mode of the complex where the six ligands move in unison away and toward the central atom that remains stationary. This is referred to as the *breathing mode*.

Using the same data for Rr_1 in Table 7.2 with the characters for the E_g and T_{1u} irreducible representations in (7.2) gives

$$S(E_g) = N \sum_R \chi_{E_g}(R)Rr_1 = \frac{1}{2\sqrt{3}}(2r_1 - r_2 - r_3 - r_4 - r_5 + 2r_6) \tag{7.4}$$

and

$$S(T_{1u}) = N \sum_R \chi_{T_{1u}}(R)Rr_1 = \frac{1}{\sqrt{2}}(r_1 - r_6). \tag{7.5}$$

In the first of these, the two atoms along the z-axis move away from the central atom while the four atoms in the xy plane all move together toward the central ion. In the second of these, the two atoms along the z-axis move up and down together while the atoms in the xy plane remain in their equilibrium positions. The three normal modes of vibration derived here are shown in Fig. 7.2.

The same procedure is done for each of the r_i and θ_{ij} coordinates. As an example of using one of the angular coordinates in (7.2), consider θ_{15} as shown in Fig. 7.1b. The transformation properties for this angle under the symmetry operations of the O_h point group are shown in Table 6.3. The C_4^2 (C_2) rotation about the z-axis takes θ_{15} into θ_{13} so a 1 appears in the θ_{13} column for this transformation row.

Table 7.3 Transformation of the θ_{15} coordinate in O_h symmetry

$R\theta_{15}$											
R	θ_{12}	θ_{13}	θ_{14}	θ_{15}	θ_{23}	θ_{34}	θ_{45}	θ_{52}	θ_{62}	θ_{63}	θ_{64}	θ_{65}
E	0	0	0	1	0	0	0	0	0	0	0	0
$6C_4$	1	1	1	0	0	0	1	1	0	0	0	1
$3C_4^2$	0	1	0	0	0	0	0	0	0	1	0	1
$6C_2$	0	0	0	1	1	0	0	1	1	1	1	0
$8C_3$	1	0	1	0	1	1	1	1	1	0	1	0
i	0	0	0	0	0	0	0	0	0	1	0	0
$6S_4$	0	1	0	0	1	1	0	0	1	0	1	1
$8S_6$	1	0	1	0	1	1	1	1	1	0	1	0
$3\sigma_h$	0	1	0	1	0	0	0	0	0	0	0	1
$6\sigma_d$	1	0	1	1	0	1	1	0	0	1	0	0

The twofold rotation about the x-axis takes θ_{15} into θ_{13} while the C_2 operation about the y-axis takes θ_{15} into θ_{65}. This analysis for all of the transformation elements gives the entries in Table 7.3. The normal vibrational modes can then be found using these data in (7.2) along with the characters of the irreducible representations of Γ_θ from Table 6.1. As an example, consider the T_{2g} irreducible representation. Equation (7.2) becomes

$$S(T_{2g}) = N\sum_R \chi_{T_{2g}}(R)R\theta_{15} = N4(-\theta_{13} + \theta_{15} + \theta_{63} - \theta_{65}),$$

which can be normalized to give

$$S(T_{2g}) = \frac{1}{2}(\theta_{15} + \theta_{63} - \theta_{13} - \theta_{65}). \qquad (7.6)$$

This describes a vibrational mode in which atoms 1 and 5 are moving away from each other, atoms 6 and 3 are moving away from each other, and atoms 0, 2, and 4 remain stationary.

The results of this type of analysis for all of the symmetry coordinates are summarized in Fig. 7.2. For each of the six types of degenerate symmetry modes an example of the atomic motion is shown. Of the 15 normal modes of vibration, six are even parity and nine are odd parity functions. They are divided into one nondegenerate symmetry mode, one doubly degenerate symmetry mode, and four triply degenerate symmetry modes.

This example shows how local vibrational modes can be designated in terms of the irreducible representations of the point symmetry group of the local complex. Each of these vibrational modes is associated with an energy level of the system. Transitions between these energy levels are discussed below. In Sect. 7.2 nonlocalized vibrational modes are treated.

7.2 Symmetry and Lattice Vibrational Modes

The vibrational characteristics of the extended lattice of atoms can be characterized in terms of a set of normal modes of vibration as was done in Sect. 7.1 for a local mode complex of atoms. In this case, translational symmetry is important and the phonon momentum or wave vector must be considered. The motion of the atoms in one unit cell of the lattice is analyzed and two situations are distinguished. In the first case, the atoms in one unit cell move together with respect to atoms in neighboring unit cells. This type of vibration is termed an *acoustic mode*. In the other case, the atoms in a given unit cell move with respect to each other. This type of vibration is termed an *optic mode*. As in Sect. 7.1, each normal mode of vibration can be associated with an irreducible representation of the symmetry group of the crystal. However, in this case the relevant symmetry group is determined by how the phonon wave vector q transforms.

The solution of the vibrational analysis problem for a crystal lattice shows that a lattice phonon can be described as a wave traveling through the lattice. In analogy with other quasiparticles in a periodic lattice (see Chap. 8), this is expressed in terms of a Block wave function [4]

$$f(\vec{R}) = \vec{u}_q(\vec{R})e^{i\vec{q}\cdot\vec{R}}, \tag{7.7}$$

where $\vec{u}_q(\vec{R})$ is the displacement of the atom at position $\vec{R}$ from its equilibrium position for a phonon of wave vector $\vec{q}$. These functions form the basis functions for the space group of the crystal lattice. As was done in Sect. 7.1, it is best to illustrate this with an example.

The crystal structure of strontium titanate ($SrTiO_3$) at room temperature is cubic perovskite belonging to the space group $O_h{}^1$. The unit cell for this crystal is shown in Fig. 7.3a. It has one molecule per unit cell and the positions of the five atoms of this molecule are shown in the figure. Since there are 15 internal degrees of freedom there will be 15 optic normal modes of vibration. In addition the atoms of the unit cell will move together to give three acoustic modes of vibration. The first Brillouin zone for this crystal structure is shown in Fig. 7.3b. There are seven points of special symmetry for the q vector in this Brillouin zone as shown in the figure. To characterize all of the normal modes of vibration of $SrTiO_3$ it is necessary to analyze each of these seven points plus a generic position of q within the Brillouin zone [2, 3, 5, 6]. The special points generally make the greatest contributions to the phonon density of states of the material. Thus they play an important role in its thermal properties.

O_h^1 is a symmorphic space group so it is possible to factor the translation and rotation symmetry operations and treat them separately. All symmetry operations that leave q invariant or transform it into $q+Q$, where Q is a primitive vector of reciprocal space, form a symmetry group called the *group of the wave vector*. This is designated $G_o(q)$ for each point q in the Brillouin zone. The irreducible representations of this group are used to designate the phonon modes at that point in the Brillouin zone. A pure rotation operation in $G_o(q)$ represented by C operating on the

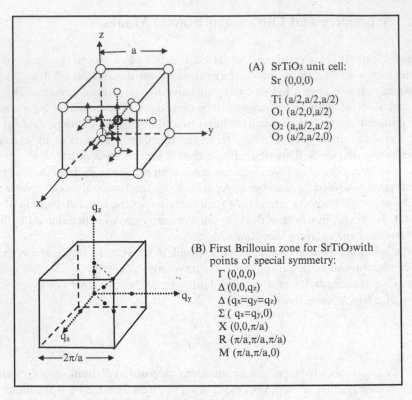

(A) SrTiO₃ unit cell:
Sr (0,0,0)

Ti (a/2,a/2,a/2)
O₁ (a/2,0,a/2)
O₂ (a,a/2,a/2)
O₃ (a/2,a/2,0)

(B) First Brillouin zone for SrTiO₃ with
points of special symmetry:
Γ (0,0,0)
Δ (0,0,q_z)
Λ (q_x=q_y=q_z)
Σ (q_x=q_y,0)
X (0,0,π/a)
R ($\pi/a,\pi/a,\pi/a$)
M ($\pi/a,\pi/a$,0)

Fig. 7.3 Unit cell and first Brillouin zone for SrTiO₃

basis function given in (7.7) will leave this function invariant or transform it into another member of a set of basis function representing a degenerate mode of vibration. All members of the set will have the same **q** vector. This is expressed as

$$C\left\{u_q\left(\vec{R}\right)e^{i\vec{q}\cdot\vec{R}}\right\} = \left\{Cu_q\left(\vec{R}\right)\right\}e^{i\vec{q}\cdot\vec{CR}} = u_q\left(\vec{R}'\right)e^{iC\vec{q}\cdot\vec{R}}$$

$$= u_q\left(\vec{R}'\right)e^{i\vec{q}\cdot\vec{R}}e^{i\vec{Q}\cdot\vec{R}}. \tag{7.8}$$

The character of the symmetry operator C will be the trace of the transformation matrix multiplied by the factor $e^{i\vec{Q}\cdot\vec{R}}$. As demonstrated in Sect. 7.1 for a complex of atoms, only those atoms whose position remains unchanged under operation C will appear along the diagonal of the transformation matrix and thus contribute to the trace. The character is then similar to (7.1) except that the factor of the number of ions remaining unchanged is replaced by the sum over the vector positions of these atoms in the factor $e^{i\vec{Q}\cdot\vec{R}}$

$$\chi_C(\alpha) = \sum_u (\pm 1 + 2\cos\alpha)e^{i\vec{Q}\cdot\vec{R}_u}. \tag{7.9}$$

As usual the $+$ and $-$ signs refer to proper and improper rotations. This expression can be used to find the characters of the symmetry operators at each point in the Brillouin zone. The resulting representations $\Gamma_v(\mathbf{q})$ can be reduced in terms of the irreducible representations of the group of the $\mathbf{q}$ vector at that point. The phonon modes at that point in the Brillouin zone are represented by these irreducible representations.

First consider the Γ point at the center of the Brillouin zone. For this location $\mathbf{Q}=0$ and the group of the q vector is the O_h point group. Applying (7.9) to each of the symmetry elements of this group gives the characters shown in Table 7.4. For this case the sum in (7.9) just gives the number of atoms that remain unchanged under the symmetry transformation or move the atom to an equivalent position in a neighboring unit cell. Multiplying this number by the rotational factor in (7.9) gives the character of the vibrational representation $\Gamma_v(\Gamma)$ at the center of the Brillouin zone. As shown in Table 7.4 this is a reducible representation that can be reduced in terms of $4T_{1u}$ and $2T_{2u}$ irreducible representations of the O_h point group. Since each of these irreducible representations is triply degenerate, the vibrational motion has 18 normal modes. Three of these are acoustic modes where the ions in the unit cell move together. This type of motion transforms as a vector which in O_h symmetry forms the basis for a T_{1u} irreducible representation. The other three triply degenerate T_{1u} representations and the two triply degenerate T_{2u} representations describe the 15 optic vibrational modes.

Next consider the point R at the corner of the Brillouin zone as shown in Fig. 7.3b. At this point the wave vector has the dimensions $q(\pi/a,\pi/a,\pi/a)$ and $\mathbf{Q}$ is not necessarily zero. The group of the q vector $G_o(\mathbf{q})$ is made up of all symmetry operations that leave the vector $q(\pi/a,\pi/a,\pi/a)$ invariant or transform it into $q(\pi/a,\pi/a,\pi/a)+\mathbf{Q}$. For example, a rotation of 90° about the z-axis takes the q vector at point R in the Brillouin zone into the vector $q(-\pi/a,\pi/a,\pi/a)$ which is the equivalent point in the neighboring Brillouin located at $Q(-2\pi/a,0,0)$. There are five other C_4 rotation operations in this class. In addition, there are three $C_2(C_4^2)$ operations that take $q(R)$ into an equivalent point in a neighboring Brillouin zone. For the C_{2z} operation the primitive vector in reciprocal space is $Q(-2\pi/a,-2\pi/a,0)$. There are $8C_3$ rotations

Table 7.4 Vibrational representations at the center and corner of the Brillouin zone for $SrTiO_3$

$\Gamma(O_h)$	E	$8C_3$	$6C_2$	$6C_4$	$3C_4^2$	i	$6S_4$	$8S_6$	$3\sigma_h$	$6\sigma_d$	
$(\pm 1+2\cos\alpha)$	3	0	-1	1	-1	-3	-1	0	1	1	
$\sum_u e^{i\vec{Q}\cdot\vec{R}_u}$	5	2	3	3	5	5	3	2	5	3	
$\Gamma_v(\Gamma)$	15	0	-3	3	-5	-15	-3	0	5	3	$4T_{1u}+2T_{2u}$
$R(O_h)$											
$(\pm 1+2\cos\alpha)$	3	2	-1	1	-1	-3	-1	0	1	1	
$\sum_u e^{i\vec{Q}\cdot\vec{R}_u}$	5	0	1	-1	1	3	1	-2	-1	3	
$\Gamma_v(R)$	15	0	-1	-1	-1	-9	-1	0	-1	3	$T_{2g}+A_{2u}+E_u+2T_{1u}+T_{2u}$

about the diagonal axes of the cubic Brillouin zone that leave R invariant or transform it into $q(R)+\mathbf{Q}$. There are six C_2 rotations about the axes bisecting the edges of the cube and parallel to the face diagonals. In addition there is the inversion operation and the operations formed by the combination of inversion with the operations already mentioned. All of these together plus the identity operation give $G_o(\mathbf{R})=O_h$. Since this is the same symmetry as found at the center of the Brillouin zone, the vibrational analysis at point R is summarized in Table 7.4.

In order to find the characters of the vibrational representation for point R shown in Table 7.4, the summation factor in (7.9) has to be calculated for each operation. The results of doing this are summarized in Table 7.5, for example, operations of each class of O_h. For each operator the vector positions for the atoms whose positions remain unchanged by the operation are listed and the value of the reciprocal lattice vector $\mathbf{Q}$ at point R for that operation is listed. Using these two quantities in the exponential factor of (7.9) along with the factor for the normal character of the operation, the summation can be evaluated to give the character of the vibrational representation $\Gamma_v(R)$ given in Table 7.4. This irreducible representation can be reduced in terms of the irreducible representations of the O_h symmetry group to give the normal modes of vibration at this point in the Brillouin zone. These 15 normal modes are divided up into four triply degenerate modes, one of which is even parity and the other three odd parity, one doubly degenerate odd parity mode, and one nondegenerate odd parity mode.

The other two points on the surface of the Brillouin zone can be analyzed in the same way. The X point is at the center of the top face of the cube with $\mathbf{q}(0,0,\pi/a)$. The group of the q vector $G_o(\mathbf{X})$ contains some of the same symmetry elements as $G_o(\mathbf{R})$. However neither the C_3 nor the S_6 classes are present. In addition, only the two C_4 and S_4 operations about the z-axis leave this q invariant so the other four operations in these classes are not part of the group. The C_{2z} rotation forms a class by itself and C_{2x} and C_{2y} form a class of operations. Two of the six diagonal C_2 axes are present. One of the three horizontal mirror planes remains as do two of the diagonal mirror planes. In addition there is a σ_v mirror plane that contains $q(X)$. These 15 symmetry elements form the D_{4h} point group which is a subgroup of O_h. The same analysis for the M point at the center of one edge of the first Brillouin zone shown in Fig. 7.3b shows that the $G_o(\mathbf{M})$ is also D_{4h}. For this point the wave vector is $\mathbf{q}(\pi/a,\pi/a,0)$. Table 7.5 lists the values of $Q(X)$ and $Q(M)$ for some of the operations of D_{4h} along with the summation factor in (7.9) and the characters for these operators in the reducible representations $\Gamma_v(X)$ and $\Gamma_v(M)$. These are used in Table 7.6 along with the characters of the irreducible representations of the D_{4h} group to determine the symmetry designations of the normal modes of vibration at points X and M on the surface of the Brillouin zone. At each point there are 15 vibrational modes divided into singly and doubly degenerate representations. These are divided into even parity and odd parity modes.

The normal modes of vibration at the three points of special symmetry within the first Brillouin zone, Δ, Λ, and Σ, can be found using the same type of analysis described above. However, $G_o(\mathbf{q})$ for these points is a subgroup of the O_h group so a simpler procedure for finding the normal vibrational modes is to use the compatibility

Table 7.5 Characters for symmetry operations at different points on the surface of the first Brillouin zone

Operator	$(\pm 1 + 2\cos\alpha)$	Position vectors $\mathbf{R}$ for unchanged atoms	Reciprocal lattice vector $\mathbf{Q}(q)$	$\sum_u e^{i\vec{Q}\cdot\vec{R}_u}$	$\chi(\alpha)$
C_{4z}	1	$\mathbf{R}(Sr)=(0,0,0)$	$\mathbf{Q}(R)=(-2\pi/a,0,0)$	-1	-1
		$\mathbf{R}(Ti)=(a/2,a/2,a/2)$	$\mathbf{Q}(X)=(0,0,0)$	3	3
		$\mathbf{R}(O_3)=(a/2,a/2,0)$	$\mathbf{Q}(M)=(-2\pi/a,0,0)$	-1	-1
C_{2z}	-1	$\mathbf{R}(Sr)=(0,0,0)$	$\mathbf{Q}(R)=(-2\pi/a,-2\pi/a,0)$	1	-1
		$\mathbf{R}(Ti)=(a/2,a/2,a/2)$	$\mathbf{Q}(X)=(0,0,0)$	5	-5
		$\mathbf{R}(O_1)=(a/2,0,a/2)$	$\mathbf{Q}(M)=(-2\pi/a,-2\pi/a,0)$	1	-1
		$\mathbf{R}(O_2)=(0,a/2,a/2)$			
		$\mathbf{R}(O_3)=(a/2,a/2,0)$			
C_2' (45° x,y)	-1	$\mathbf{R}(Sr)=(0,0,0)$	$\mathbf{Q}(R)=(0,0,-2\pi/a)$	1	-1
		$\mathbf{R}(Ti)=(a/2,a/2,a/2)$	$\mathbf{Q}(X)=(0,0,-\pi/2)$	1	-1
		$\mathbf{R}(O_3)=(a/2,a/2,0)$	$\mathbf{Q}(M)=(0,0,0)$	3	-3
C_3 (0− xyz)	0	$\mathbf{R}(Sr)=(0,0,0)$	$\mathbf{Q}(R)=(0,0,0)$	2	0
		$\mathbf{R}(Ti)=(a/2,a/2,a/2)$			
i	-3	$\mathbf{R}(Sr)=(0,0,0)$	$\mathbf{Q}(R)=(-2\pi/a,-2\pi/a,-2\pi/a)$	3	-9
		$\mathbf{R}(Ti)=(a/2,a/2,a/2)$	$\mathbf{Q}(X)=(0,0,-2\pi/a)$	-1	3
		$\mathbf{R}(O_1)=(a/2,0,a/2)$	$\mathbf{Q}(M)=(-2\pi/a,-2\pi/a,0)$		
		$\mathbf{R}(O_2)=(0,a/2,a/2)$			
		$\mathbf{R}(O_3)=(a/2,a/2,0)$			
S_{4z}	-1	$\mathbf{R}(Sr)=(0,0,0)$	$\mathbf{Q}(R)=(-2\pi/a,0,-2\pi/a)$	1	-1
		$\mathbf{R}(Ti)=(a/2,a/2,a/2)$	$\mathbf{Q}(X)=(0,0,-2\pi/a)$	1	-1
		$\mathbf{R}(O_3)=(a/2,a/2,0)$	$\mathbf{Q}(M)=(0,-2\pi/a,0)$	-1	1
S_6 (0− xyz)	-2	$\mathbf{R}(Sr)=(0,0,0)$	$\mathbf{Q}(R)=(-2\pi/a,-2\pi/a,-2\pi/a)$	0	0
		$\mathbf{R}(Ti)=(a/2,a/2,a/2)$			
$\sigma_h(xy)$	1	$\mathbf{R}(Sr)=(0,0,0)$	$\mathbf{Q}(R)=(0,0,-2\pi/a)$	-1	-1
		$\mathbf{R}(Ti)=(a/2,a/2,a/2)$	$\mathbf{Q}(X)=(0,0,-2\pi/a)$		
		$\mathbf{R}(O_1)=(a/2,0,a/2)$	$\mathbf{Q}(M)=(0,0,0)$	5	5
		$\mathbf{R}(O_2)=(0,a/2,a/2)$			
		$\mathbf{R}(O_3)=(a/2,a/2,0)$			
σ_d (45° x,y)	1	$\mathbf{R}(Sr)=(0,0,0)$	$\mathbf{Q}(R)=(0,0,0)$	3	3
		$\mathbf{R}(Ti)=(a/2,a/2,a/2)$	$\mathbf{Q}(X)=(0,0,0)$		
		$\mathbf{R}(O_3)=(a/2,a/2,0)$	$\mathbf{Q}(M)=(0,0,0)$	3	3
E	3	$\mathbf{R}(Sr)=(0,0,0)$	$\mathbf{Q}(R)=(0,00)$	5	15
		$\mathbf{R}(Ti)=(a/2,a/2,a/2)$	$\mathbf{Q}(X)=(0,0,0)$	5	15
		$\mathbf{R}(O_1)=(a/2,0,a/2)$	$\mathbf{Q}(M)=(0,0,0)$		
		$\mathbf{R}(O_2)=(0,a/2,a/2)$			
		$\mathbf{R}(O_3)=(a/2,a/2,0)$			

Table 7.6 Vibrational representations at the X and M points of the Brillouin zone for SrTiO$_3$

D_{4h}	E	$2C_4$	C_2	$2C_2'$	$2C_2''$	i	$2S_4$	σ_h	$2\sigma_v$	$2\sigma_d$	
A_{1g}	1	1	1	1	1	1	1	1	1	1	
A_{2g}	1	1	1	-1	-1	1	1	1	-1	-1	
B_{1g}	1	-1	1	1	-1	1	-1	1	1	-1	
B_{2g}	1	-1	1	-1	1	1	-1	1	-1	1	
E_g	2	0	-2	0	0	2	0	-2	0	0	
A_{1u}	1	1	1	1	1	-1	-1	-1	-1	-1	
A_{2u}	1	1	1	-1	-1	-1	-1	-1	1	1	
B_{1u}	1	-1	1	1	-1	-1	1	-1	-1	1	
B_{2u}	1	-1	1	-1	1	-1	1	-1	1	-1	
E_u	2	0	-2	0	0	-2	0	2	0	0	
$X(D_{4h})$											
$(\pm 1 + 2\cos\alpha)$	3	1	-1	-1	-1	-3	-1	1	1	1	
$\sum_u e^{iQ\cdot R_u}$	5	3	5	-1	1	-1	1	-1	5	3	
$\Gamma_v(X)$	15	3	-5	1	-1	3	-1	-1	5	3	$2A_{1g}+B_{1g}+3E_g+2A_{2u}+2E_u$
$M(D_{4h})$											
$(\pm 1 + 2\cos\alpha)$	3	1	-1	-1	-1	-3	-1	1	1	1	
$\sum_u e^{iQ\cdot R_u}$	5	-1	1	-1	3	1	-1	5	-1	3	
$\Gamma_v(R)$	15	-1	-1	1	-3	-3	1	5	-1	3	$A_{1g}+A_{2g}+B_{1g}+B_{2g}+E_g+A_{2u}+2B_{1u}+3E_u$

relationships between the irreducible representations of a group and its subgroups. As discussed in Chap. 7, the irreducible representations of a group transform as reducible representations in its subgroups. Thus the T_{1u} and T_{2u} irreducible representations of O_h can be reduced in terms of the irreducible representations of $G_o(\Delta)$, $G_o(\Lambda)$, and $G_o(\Sigma)$. For the Δ points along the z-axis, $\mathbf{q}(0,0,q_z)$ transforms into itself under the C_{4z} and C_{2z} rotational operations as well as the two σ_v and two σ_d mirror planes that contain the z-axis. These elements form the C_{4v} point group that is a subgroup of O_h. For the Λ points along the cube diagonal axis, $\mathbf{q}(q_x=q_y=q_z)$ transforms into itself under the C_3 rotational operations that contain this diagonal as well as the three σ_v mirror. These elements form the C_{3v} point group that is a subgroup of O_h. For the Σ points along the diagonal axis in the xy plane, $\mathbf{q}(q_x=q_y,0)$ transforms into itself under the C_{2z} rotational operation and the two mirror planes that contain this q vector axis. These elements form the C_{2v} point group that is a subgroup of O_h. A generic point in the first Brillouin zone has a q vector that only transforms into itself under the identity transformation which is described by the C_1 point group.

Using the concepts discussed in Chap. 2, an example of determining the correlation between the irreducible representations of O_h and its C_{4v} subgroup is shown in Table 7.7. This shows that the four T_{1u} normal modes of vibration of O_h symmetry at the center of the Brillouin zone become four A_1 and four E modes of vibration in C_{4v} symmetry as the wave vector moves along the z-axis of the Brillouin zone. The triply degenerate T_{2u} modes at Γ become a nondegenerate B_1 mode and a doubly degenerate E mode in the region of C_{4v} symmetry.

A summary of the correlations of representations of O_h and its C_1, C_{2v}, C_{3v}, and C_{4v} subgroups is shown in Fig. 7.4. In addition, the correlations of the representations of the D_{4h} group and its C_{2v} and C_{4v} subgroups are shown. This gives all of the symmetry designations of the 15 normal modes of vibration at every point in the

Table 7.7 Correlation table for O_h and C_{4v} point groups

O_h	E	$8C_3$	$6C_2$	$6C_4$	$3C_4^2$	i	$6S_4$	$8S_6$	$3\sigma_h$	$6\sigma_d$	Correlation with C_{4v}
A_{1g}	1	1	1	1	1	1	1	1	1	1	A_1
A_{2g}	1	1	−1	−1	1	1	−1	1	1	−1	B_1
E_g	2	−1	0	0	2	2	0	−1	2	0	A_1+B_1
T_{1g}	3	0	−1	1	−1	3	1	0	−1	−1	A_2+E
T_{2g}	3	0	1	−1	−1	3	−1	0	−1	1	B_2+E
A_{1u}	1	1	1	1	1	−1	−1	−1	−1	−1	A_2
A_{2u}	1	1	−1	−1	1	−1	1	−1	−1	1	B_2
E_u	2	−1	0	0	2	−2	0	1	−2	0	A_2+B_2
T_{1u}	3	0	−1	1	−1	−3	−1	0	1	1	A_1+E
T_{2u}	3	0	1	−1	−1	−3	1	0	1	−1	B_1+E

C_{4v}	E	$2C_4$	C_2	$2\sigma_v$	$2\sigma_d$
A_1	1	1	1	1	1
A_2	1	1	1	−1	−1
B_1	1	−1	1	1	−1
B_2	1	−1	1	−1	1
E	2	0	−2	0	0

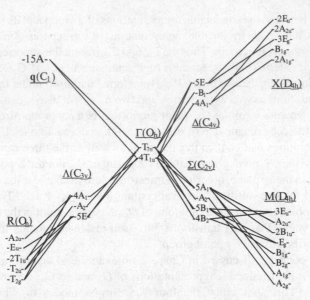

Fig. 7.4 Symmetry designations of the 15 normal modes of vibration of SrTiO$_3$ at each point in the first Brillouin zone

first Brillouin zone of SrTiO$_3$. Note that the subgroup analysis is consistent with the full vibrational analysis at the X, M, and R points described above.

The variation of the phonon frequency as a function of wave vector is called *phonon dispersion*. The dispersion of each of the different phonon modes can be measured experimentally by techniques such as neutron scattering. The dispersion curves vary smoothly throughout the first Brillouin zone with different shapes depending on the wave vector direction. The slope of the dispersion curve is proportional to the phonon velocity and goes to zero at the zone boundary. Section 7.3 describes transitions between these vibrational states of the system caused by the absorption, emission, or scattering of light.

7.3 Transitions Between Vibrational Energy Levels

The normal modes of vibration discussed in Sect. 7.2 are associated with phonons of quantized energy and momentum. Phonons have very low energies and the absorption or emission of phonons can cause transitions to occur between the low energy electronic energy levels of the material. If no photons are involved, these are called *radiationless processes*. If low energy infrared photons are simultaneously absorbed or emitted with the creation or annihilation of phonons, these are radiative process similar to those discussed in Chap. 4.

7.3.1 Radiationless Transitions

The Hamiltonian involved in radiationless transitions is the electron–phonon interaction Hamiltonian H_{ep}. The source of this interaction is that the atomic vibrations associated with a specific phonon mode modulate the local crystal field and thus the electronic energy levels of the system (as discussed in Chap. 4). The interaction Hamiltonian describing this effect can be expressed in terms of an expansion of the crystal field modulation as

$$H_{ep} = \sum_q V_q S_q + \cdots, \qquad (7.10)$$

where the V_q represents the electron–phonon coupling coefficient, which is the derivative of the crystal field with respect to the normal vibrational mode coordinate S_q. Here the parameter q designates a specific phonon including its branch, frequency, wave vector, and polarization. For the absorption or emission of a single phonon in a transition only the first term in the expansion of (7.10) is needed.

In the basic treatment of lattice vibrations, the phonons are modeled as an ensemble of linear harmonic oscillators [2, 4]. The quantum mechanical solution to Schrödinger's equation for harmonic oscillators shows that the wavefunctions for the vibrational energy levels are expressed as a series of Hermite polynomials of order n, where n is the vibrational quantum number for the system. For a specific normal mode q,

$$\psi_n(q) = N e^{(1/2)\alpha^2 q^2} H_n(\alpha q). \qquad (7.11)$$

Here N is a normalization constant and $\alpha = \sqrt{\omega/\hbar}$. The first several values of the Hermite polynomials are

$$H_0(x) = 1, H_1(x) = 2x, H_2(x) = 4x^2 - 2, \ldots.$$

The quantum number n designates the degree of excitation of the q normal mode with $n=0$ being the ground state, $n=1$ the first excited state, etc. The vibrational wave functions for the entire system are products of these single mode wavefunctions.

Using second quantized notation, the vibrational wave functions are expressed in terms of occupation numbers of each type of normal mode. For treating radiationless processes, the wave function for the system can be written as a product of the electronic and vibrational parts

$$\Psi = |\psi_{el}\rangle |n_1\rangle |n_2\rangle \cdots |n_q\rangle \cdots, \qquad (7.12)$$

where the n_q are the occupation numbers of specific phonon modes.

The normal mode coordinate S_q can then be expressed in terms of phonon, their creation and annihilation operators

$$S_q = \sqrt{\frac{\hbar}{2\omega_q}}\left(b_q + b^*_{-q}\right). \tag{7.13}$$

The operator b_q annihilates a normal vibrational mode designated by q while the operator b^* creates a phonon designated by $-q$. The electron–phonon interaction expressed in this formalism is

$$H_{ep} = \sum_q V_q \sqrt{\frac{\hbar}{2\omega_q}}\left(b_q + b^*_{-q}\right). \tag{7.14}$$

Using the wave functions from (7.12) and the interaction Hamiltonian from (7.11) in the expression for the transition rate from time-dependent perturbation theory gives

$$w_{nr} = \frac{2\pi}{\hbar}\left|\langle \Psi_f|H_{ep}|\Psi_i\rangle\right|^2 \rho_f, \tag{7.15}$$

where Ψ_i and Ψ_f represent the initial and final states of the transition and ρ_f is the density of final states. For a transition involving absorption of one phonon of the type designated by q the transition rate is given by

$$
\begin{aligned}
w_{nr}^{ab} &= \frac{2\pi}{\hbar}\left(\frac{\hbar\omega_q}{2Mv^2}\right)\left|\langle \psi_f^{el}|V_q|\psi_i^{el}\rangle\right|^2 \left|\langle n_q - 1|b_q|n_q\rangle\right|^2 \rho_f \\
&= \left(\frac{3\omega_q^3}{2\pi\rho v^5 \hbar}\right) n_q \left|\langle \psi_f^{el}|V_q|\psi_i^{el}\rangle\right|^2.
\end{aligned}
\tag{7.16}
$$

For a transition involving the emission of one phonon of the type designated by q the transition rate is given by

$$w_{em}^{em} = \left(\frac{3\omega_q^3}{2\pi\rho v^5 \hbar}\right)(n_q + 1)\left|\langle \psi_f^{el}|V_q|\psi_i^{el}\rangle\right|^2. \tag{7.17}$$

These two expressions for transition rates have made use of the properties of the phonon creation and annihilation operators [2]

$$b_q|n_q\rangle = \sqrt{n_q}|n_q - 1\rangle, \quad b_q^*|n_q\rangle = \sqrt{n_q + 1}|n_q + 1\rangle. \tag{7.18}$$

Also the material density has been defined as $\rho = M/V$. The density of final states in (7.15) is the product of a delta function for the electronic transition energy and the

Debye phonon density of states [4]. The former function has a factor of $(1/\hbar)$ needed to convert from energy to frequency units.

The selection rules that determine whether the transitions described by (7.17) and (7.18) are allowed or forbidden are contained in the matrix elements of the electron–phonon interaction operator between the initial and final electronic states of the system. In Chap. 4 it was shown that the electronic states of the system transform as the basis functions for irreducible representations of the relevant symmetry group. Since V_q is the derivative of the crystal field with respect to the specific phonon symmetry coordinate, the electron–phonon interaction operator transforms as irreducible representation associated with the q normal mode of vibration. Thus, in terms of group theory, the matrix elements in (7.17) and (7.18) are expressed as

$$\langle \psi_f^{el} | V_q | \psi_i^{el} \rangle = \Gamma_f \times \Gamma_q \times \Gamma_i \supset A_{1g} \text{ for an allowed transition.}$$

Since the totally symmetric irreducible representation only appears in the reduction of a direct product of a representation with itself, this expression can be rewritten as

$$\Gamma_f \times \Gamma_i \supset \Gamma_q. \qquad (7.19)$$

The character tables of the different point groups and the techniques of forming and reducing direct product representations discussed in Chap. 2 can be used with (7.19) to determine the normal mode symmetries causing allowed transitions between a specific set of electronic states. As an example of this, consider the $SrTiO_3$ structure described in Sect. 7.2. If this crystal has electronic states represented by E_u and A_{2g} irreducible representations in O_h symmetry their direct product reduces to the E_u irreducible representation. Since there are no normal modes of vibration transforming as E_u at the center of the Brillouin zone, none of the Γ point phonons contribute to radiationless transitions between these electronic energy levels. However, there is an E_u phonon with wave vector $q(\pi/a,\pi/a,\pi/a)$ at the R point of the Brillouin zone that is allowed for this transition. If these electronic energy levels transform as E_u and A_{2g} in the D_{4h} symmetry group, then the phonons with E_u symmetry at the X and M points in the Brillouin zone can contribute to allowed radiationless transitions. For electronic energy levels transforming as T_{1u} and E_g in O_h symmetry, the reduction of their direct product contains the T_{1u} and T_{2u} irreducible representations. Thus all 15 vibrational modes at the Γ point and three of the triply degenerate modes at the R point can contribute to allowed radiationless transitions between these levels.

7.3.2 Infrared Transitions

It is also possible to have transitions between different vibrational energy levels of the system accompanied by the absorption or emission of photons. Since the difference in energy levels is small, the photons resonant with the transition have low energies and appear in the infrared region of the electromagnetic spectrum.

In the case of the absorption of an infrared photon, one or more phonons are created. This process must conserve both energy and momentum. These requirements are expressed as

$$\omega = ck \qquad\qquad (7.20)$$

and

$$\vec{k} = \vec{q} + \vec{Q}. \qquad\qquad (7.20)$$

Here ω is the phonon frequency, $\vec{k}$ is the photon wave vector, $\vec{q}$ is the phonon wave vector, and $\vec{Q}$ is a reciprocal lattice vector. These conservation requirements are illustrated in Fig. 7.5. This figure shows an example of how the energy of phonons vary as a function of their momentum. Typical acoustic phonon modes have zero energy at the center of the Brillouin zone. Their energy increases with **q** and becomes flat near the surface of the first Brillouin zone. Optic modes on the other hand have some finite energy at **q**=0, decrease with increasing **q**, and become flat near the surface of the Brillouin zone. Plotting the same type of energy versus momentum curve for photons on this same scale gives a very steep straight line near the y-axis. This is because the velocity of photons is over five orders of magnitude greater than the velocity of phonons (sound). The conservation laws are satisfied at the points that the photon and phonon curves intersect on this plot. This shows that the acoustic modes of vibration do not contribute to infrared absorption and that only the optic modes at or very near the center of the Brillouin can contribute to infrared absorption.

The selection rules for infrared absorption can again be determined from the transition matrix element. Since infrared radiation is part of the electromagnetic spectrum, the interaction Hamiltonian is the electric dipole term of the radiation field similar to (4.23). However, the initial and final states are now represented by the wave functions given in (7.12). Thus,

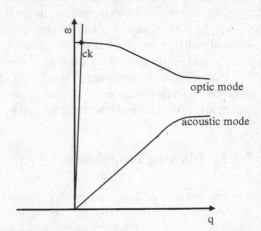

Fig. 7.5 Energy versus momentum curves for phonons and photons

$$M_{\text{ir}} = |\langle\psi_1(q)|H_{\text{ed}}|\psi_0(q)\rangle|. \tag{7.21}$$

As discussed in Chap. 4, the electric dipole operator transforms as the components of a vector. The ground state vibrational wave function transforms as the exponential factor in (7.11). This has the same form as the harmonic oscillator potential energy and thus as part of the Hamiltonian for the system it is invariant under all the operations of the symmetry of the system. Therefore $\psi_0(q)$ transforms as the totally symmetric irreducible representation $A_{1\text{g}}$. Since the first excited state $\psi_1(q)$ has the additional factor of q from the $H_1(\alpha q)$, it will transform according to the same irreducible representation as the q normal mode. The symmetry selection rules are then given by

$$\Gamma_1^q \times \Gamma_{\text{ed}} \times \Gamma_0^q \supset A_{1\text{g}}. \tag{7.22}$$

For this case $\Gamma_0^q = A_{1\text{g}}$ so the transition matrix element is nonzero only for those phonon modes transforming as one of the components of a vector. The character tables of the 32 crystallographic point groups given in Chap. 2 show the irreducible representations for which the vector components act as basis functions. Any normal vibrational mode transforming as one of these irreducible representations will be infrared active.

For the example of $SrTiO_3$ discussed in Sect. 7.2, at the center of the Brillouin zone there were 15 normal modes of vibration of which three transformed as the triply degenerate $T_{2\text{u}}$ representation and 12 transformed as four triply degenerate $T_{1\text{u}}$ modes in the O_h symmetry group. Table 2.32 shows that the three vector components of the electric dipole moment operator transform together as the triply degenerate $T_{1\text{u}}$ representation in O_h. Thus the Γ point $T_{1\text{u}}$ phonons of $SrTiO_3$ are infrared active while all other normal modes are infrared inactive.

7.4 Raman Scattering

Chapter 6 dealt with second-order nonlinear optical processes in which the energy and momentum of the incident and transmitted photons are conserved. In addition to these elastic scattering processes, photons can be inelastically scattered by a material. In this case, the transmitted photons have different energies from the incident photons with the energy difference being transferred to lattice phonons. This type of process in which phonons are created or annihilated by photons traveling through the lattice is called *Raman scattering*. It is a nonlinear optical process called *Stokes scattering* if the phonon is created so the transmitted photons have lower energy than the incident photons and is called *anti-Stokes scattering* if a phonon is annihilated so the transmitted photon has higher energy than the incident photons. Since there is no requirement for resonance with the difference in energy levels of the system as there is for infrared absorption, any wavelength of light can be used for Raman scattering. The measurement of Raman scattering, called Raman spectroscopy, has proved to be a useful tool for studying the local and optical vibrational modes of solids and in areas such as analytical chemistry and remote sensing. The physical interaction causing Raman scattering can be expressed in tensor format as described below.

As discussed in Chap. 6, the nonlinear response of a material to an electromagnetic light wave can be described through higher order terms in the expansion of the dielectric susceptibility, the refractive index, or the polarizability. Chapter 6 made use of the expansion in terms of the susceptibility tensor while in this chapter the expansion in terms of the polarizability tensor is used. The two are related through the expression given at the start of Chap. 6. The polarization of a material can be expressed as

$$P_i = p_i + \alpha_{ij} E_j + \beta_{ijk} E_j E_k + \cdots. \tag{7.23}$$

In this expression the subscripts i, j, and k are vector direction components, $\mathbf{p}$ is a permanent dipole moment in the material, and the E_i are the directional components of the electric field of the light wave. The second-order and higher order terms represent the polarization induced by the light wave. The largest contribution to this induced polarization comes from the second term where $\vec{\alpha}$ is the *polarizability tensor*.

The lattice vibrations modulate both the permanent and the light-induced contributions to the polarization. This can be expressed as a Taylor series

$$P_i = p_{0i} + \sum_n \left(\frac{\partial p_i}{\partial q_n}\right)_{q=0} q_{0n} \cos(\omega_n t) + \alpha_{0ij} E_{0j} \cos(\omega_l t)$$

$$+ \sum_n \left(\frac{\partial \alpha_{ij}}{\partial q_n}\right)_{q=0} q_{0n} \cos(\omega_n t) E_{0j} \cos(\omega_l t). \tag{7.24}$$

In this expression it has been assumed that the time dependence of the vibrating atoms is given by $q_n(t) = q_{0n}\cos(\omega_n t)$ and the time dependence of the electric field is $E_i(t) = E_{0i}\cos(\omega_l t)$.

The second term in (7.24) describes a dipole moment oscillating with the frequency of a normal vibrational mode. If an infrared light wave is generated by an electric dipole oscillating at the same frequency ($\omega_l = \omega_n$), a resonance interaction can occur leading to infrared absorption as discussed in Sect. 7.3. Thus the normal vibrational modes for which this term is not zero are the infrared active modes.

The third term describes a light-induced polarization wave oscillating at the same frequency of the incident light wave. This part of the polarization wave produces the transmitted light wave having the same frequency as the incident light wave that is called *Rayleigh scattering*.

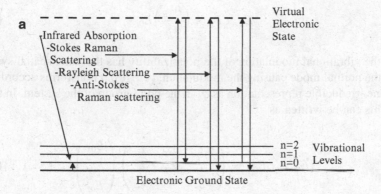

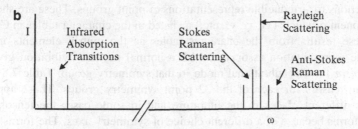

Fig. 7.6 Schematic picture of infrared absorption and Raman scattering (**a**) Transitions; (**b**) spectra

The product of the cosine functions in the fourth term can be rewritten using a trigonometric identity to be $\cos(\omega_l \pm \omega_n)t$. Therefore this term describes a light-induced polarization wave modulated by a phonon vibrational mode to oscillate with beat frequencies of $\pm\omega_n$ about the central frequency ω_l. This part of the polarization wave generates the Stokes and anti-Stokes Raman scattered transmitted wave. The phonon modes for which this term is not zero are called Raman active.

This simple treatment of the interaction of a light wave with lattice vibrations shows that infrared active modes are vibrations that cause a change in the local dipole moment while Raman active modes are vibrations that cause a change in the local polarizability. Figure 7.6 shows a schematic picture of the transitions and spectra associated with infrared and Raman active vibrational modes. Note that the Stokes and anti-Stokes Raman transitions are symmetric in energy about the central Rayleigh line and the Raman transitions appear at much higher energy in the spectrum than the infrared transitions.

As discussed in Chap. 6, the electric field vector of an optical wave transmitted through a material is proportional to the light-induced polarization in the material. For the case of Raman scattering this can be expressed as

$$E_{ti} \propto P_i \propto \left(\frac{\partial \alpha_{ij}}{\partial q_n}\right) E_{0j}. \tag{7.25}$$

Since the vibrational modulation of the polarizability has the same spatial symmetry as the normal mode causing the modulation, $(\partial \alpha_{ij}/\partial q_n)$ transforms according to the same irreducible representation as q_n in the point group of the system. In tensor form this can be written as

$$\left(E_{tx}, E_{ty}, E_{tz}\right) = \begin{pmatrix} \alpha_{xx} & \alpha_{xy} & \alpha_{xz} \\ \alpha_{yx} & \alpha_{yy} & \alpha_{yz} \\ \alpha_{zx} & \alpha_{zy} & \alpha_{zz} \end{pmatrix} \begin{pmatrix} E_{0x} \\ E_{0y} \\ E_{0z} \end{pmatrix}. \tag{7.26}$$

The polarizability tensor is a symmetric 3×3 matrix whose elements transform as basis functions for irreducible representations of point groups. These are shown as the component products (xx, xy, yz, etc.) as listed in the character tables in Chap. 2. Using these results from the character tables as the matrix elements in each irreducible representation associated with a normal mode of vibration gives the *Raman tensor* for that vibrational mode in that symmetry group. Table 7.8 shows the Raman tensors for each of the 32 point symmetry groups. These appear in numerous different places in the literature and in some cases the tensors have different forms because of a different choice of symmetry axes. The forms shown in Table 7.8 are consistent with the character tables given in Chap. 2. Note that for a crystal with a center of symmetry the Raman active modes are even parity while the infrared active modes are odd parity.

For a normal mode of vibration to be Raman active it must induce a change in the polarizability of the material. The selection rules for Raman scattering can be expressed in terms of group theory as was done for infrared absorption in (7.22)

$$\Gamma_1^q \times \Gamma_\alpha \times \Gamma_0^q \supset A_{1g}. \tag{7.27}$$

Since the ground state always transforms as the totally symmetric irreducible representation $\Gamma_0^q = A_{1g}$, and A_{1g} only appears in the reduction of the product of a representation with itself, the Raman active modes will be those that transform as one of the components of the Raman tensor. For example, if a material belonged to the octahedral O symmetry group and had normal modes of vibration transforming as A_1, E, or T_2, they would be Raman active. On the other hand if this material had normal modes transforming as A_2 or T_1 irreducible representations it would not be Raman active.

Because of the tensor nature of (7.26), the normal modes of vibration that appear in a Raman spectroscopy experiment depend on the polarization of the incident and scattered light. For example, consider a material with a C_{2v} point group symmetry that has normal modes of vibration transforming as the A_1, A_2, B_1, and B_2 irreducible representations. According to (7.27) all of these modes will be Raman active. Using (7.26) and the Raman tensors from Table 7.8 shows that a Raman scattering

Table 7.8 Raman tensors for the 32 crystallographic point groups

Crystal system	Point group	Irreducible representations of Raman active modes and their Raman tensors			
Triclinic	C_1 C_i	A A_g $\begin{pmatrix} \alpha_{xx} & \alpha_{xy} & \alpha_{xz} \\ \alpha_{xy} & \alpha_{yy} & \alpha_{yz} \\ \alpha_{xz} & \alpha_{yz} & \alpha_{zz} \end{pmatrix}$			
Monoclinic	C_2 C_s C_{2h}	A / A' / A_g $\begin{pmatrix} \alpha_{xx} & \alpha_{xy} & 0 \\ \alpha_{xy} & \alpha_{yy} & 0 \\ 0 & 0 & \alpha_{zz} \end{pmatrix}$	B / A'' / B_g $\begin{pmatrix} 0 & 0 & \alpha_{xz} \\ 0 & 0 & \alpha_{yz} \\ \alpha_{xz} & \alpha_{yz} & 0 \end{pmatrix}$		
Orthorhombic	D_2 C_{2v} D_{2h}	A / A_1 / A_g $\begin{pmatrix} \alpha_{xx} & 0 & 0 \\ 0 & \alpha_{yy} & 0 \\ 0 & 0 & \alpha_{zz} \end{pmatrix}$	B_1 / A_2 / B_{1g} $\begin{pmatrix} 0 & \alpha_{xy} & 0 \\ \alpha_{xy} & 0 & 0 \\ 0 & 0 & 0 \end{pmatrix}$	B_2 / B_1 / B_{2g} $\begin{pmatrix} 0 & 0 & \alpha_{xz} \\ 0 & 0 & 0 \\ \alpha_{xz} & 0 & 0 \end{pmatrix}$	B_3 / B_2 / B_{3g} $\begin{pmatrix} 0 & 0 & 0 \\ 0 & 0 & \alpha_{yz} \\ 0 & \alpha_{yz} & 0 \end{pmatrix}$
Tetragonal	C_4 S_4 C_{4h}	A / A / A_g $\begin{pmatrix} \alpha_{xx} & 0 & 0 \\ 0 & \alpha_{yy} & 0 \\ 0 & 0 & \alpha_{zz} \end{pmatrix}$	B / B / B_g $\begin{pmatrix} \alpha_{xx} & \alpha_{xy} & 0 \\ \alpha_{xy} & -\alpha_{xx} & 0 \\ 0 & 0 & 0 \end{pmatrix}$	E_x / E_x / E_g $\begin{pmatrix} 0 & 0 & \alpha_{xz} \\ 0 & 0 & \alpha_{yz} \\ \alpha_{xz} & \alpha_{yz} & 0 \end{pmatrix}$	E_y / E_y / E_g $\begin{pmatrix} 0 & 0 & -\alpha_{xz} \\ 0 & 0 & \alpha_{yz} \\ -\alpha_{xz} & \alpha_{yz} & 0 \end{pmatrix}$
	C_{4v} D_4	A_1 A_1	B_1 B_1	B_2 B_2	E E

(continued)

Table 7.8 (continued)

Crystal system	Point group	Raman active modes and their Raman tensors
	D_{2d} D_{4h}	A_1 / A_{1g}: $\begin{pmatrix} \alpha_{xx} & 0 & 0 \\ 0 & \alpha_{yy} & 0 \\ 0 & 0 & \alpha_{zz} \end{pmatrix}$ B_1 / B_{1g}: $\begin{pmatrix} \alpha_{xx} & 0 & 0 \\ 0 & -\alpha_{xx} & 0 \\ 0 & 0 & 0 \end{pmatrix}$ B_2 / B_{2g}: $\begin{pmatrix} 0 & \alpha_{xy} & 0 \\ \alpha_{xy} & 0 & 0 \\ 0 & 0 & 0 \end{pmatrix}$ E / E_g: $\begin{pmatrix} 0 & 0 & \alpha_{xz} \\ 0 & 0 & 0 \\ \alpha_{xz} & 0 & 0 \end{pmatrix}$ E / E_g: $\begin{pmatrix} 0 & 0 & 0 \\ 0 & 0 & \alpha_{yz} \\ 0 & \alpha_{yz} & 0 \end{pmatrix}$
Trigonal	C_3 S	A / A_g: $\begin{pmatrix} \alpha_{xx} & 0 & 0 \\ 0 & \alpha_{yy} & 0 \\ 0 & 0 & \alpha_{zz} \end{pmatrix}$ E / E_g: $\begin{pmatrix} \alpha_{xx} & \alpha_{xy} & \alpha_{xz} \\ \alpha_{xy} & -\alpha_{xx} & \alpha_{yz} \\ \alpha_{xz} & \alpha_{yz} & 0 \end{pmatrix}$ E / E_g: $\begin{pmatrix} -\alpha_{xy} & -\alpha_{xx} & -\alpha_{xz} \\ -\alpha_{xx} & \alpha_{xy} & \alpha_{yz} \\ -\alpha_{xz} & \alpha_{yz} & 0 \end{pmatrix}$
	D_3 C_{3v} D_{3d}	A_1 / A_1 / A_{1g}: $\begin{pmatrix} \alpha_{xx} & 0 & 0 \\ 0 & \alpha_{yy} & 0 \\ 0 & 0 & \alpha_{zz} \end{pmatrix}$ E / E / E_g: $\begin{pmatrix} \alpha_{xx} & 0 & 0 \\ 0 & -\alpha_{xx} & \alpha_{yz} \\ 0 & \alpha_{yz} & 0 \end{pmatrix}$ E / E / E_g: $\begin{pmatrix} 0 & -\alpha_{xx} & -\alpha_{xy} \\ -\alpha_{xx} & 0 & \alpha_{yz} \\ -\alpha_{xy} & \alpha_{yz} & 0 \end{pmatrix}$
Hexagonal	C_6 C_{3h} C_{6h}	A / A' / A_g E_1 / E_1'' / E_{1g} E_2 / E_2' / E_{2g}

(continued)

Table 7.8 (continued)

Crystal system	Point group	Raman active modes and their Raman tensors

$$\begin{pmatrix} \alpha_{xx} & 0 & 0 \\ 0 & \alpha_{yy} & 0 \\ 0 & 0 & \alpha_{zz} \end{pmatrix} \qquad \begin{pmatrix} 0 & 0 & \alpha_{xz} \\ 0 & 0 & \alpha_{yz} \\ \alpha_{xz} & \alpha_{yz} & 0 \end{pmatrix} \qquad \begin{pmatrix} \alpha_{xx} & \alpha_{xy} & 0 \\ \alpha_{xy} & -\alpha_{xx} & 0 \\ 0 & 0 & 0 \end{pmatrix}$$

$$\begin{pmatrix} 0 & 0 & \alpha_{xz} \\ 0 & 0 & \alpha_{yz} \\ -\alpha_{xz} & \alpha_{yz} & 0 \end{pmatrix} \qquad \begin{pmatrix} \alpha_{xx} & -\alpha_{xy} & 0 \\ -\alpha_{xy} & -\alpha_{xx} & 0 \\ 0 & 0 & 0 \end{pmatrix}$$

D_6 C_{6v} D_{3h} D_{6h}

D_6	C_{6v}	D_{3h}	D_{6h}
A_1	A_1	A'	A_{1g}
E_1	E_1	E''	E_{1g}
E_1	E_1	E''	E_{1g}
E_2	E_2	E'	E_{2g}
E_2	E_2	E'	E_{2g}

A_{1g}:
$$\begin{pmatrix} \alpha_{xx} & 0 & 0 \\ 0 & \alpha_{yy} & 0 \\ 0 & 0 & \alpha_{zz} \end{pmatrix}$$

E_{1g}:
$$\begin{pmatrix} 0 & 0 & 0 \\ 0 & 0 & \alpha_{yz} \\ 0 & \alpha_{yz} & 0 \end{pmatrix} \qquad \begin{pmatrix} 0 & 0 & -\alpha_{xz} \\ 0 & 0 & 0 \\ -\alpha_{xz} & 0 & 0 \end{pmatrix}$$

E_{2g}:
$$\begin{pmatrix} 0 & \alpha_{xy} & 0 \\ \alpha_{xy} & 0 & 0 \\ 0 & 0 & 0 \end{pmatrix} \qquad \begin{pmatrix} \alpha_{xx} & 0 & 0 \\ 0 & -\alpha_{xx} & 0 \\ 0 & 0 & 0 \end{pmatrix}$$

(continued)

Table 7.8 (continued)

Crystal system	Point group	Irreducible representations of Raman active modes and their Raman tensors		
Cubic	T	A	E	T
	T_h	A_g	E_g	T_g
	T_d	A_1	E	T_2
	O	A_1	E	T_2
	O_h	A_{1g}	E_g	T_{2g}

$A,\ A_g,\ A_1,\ A_1,\ A_{1g}$:
$$\begin{pmatrix} \alpha_{xx} & 0 & 0 \\ 0 & \alpha_{yy} & 0 \\ 0 & 0 & \alpha_{zz} \end{pmatrix}$$

$E,\ E_g,\ E,\ E,\ E_g$:
$$\begin{pmatrix} \alpha_{xx} & 0 & 0 \\ 0 & \alpha_{xx} & 0 \\ 0 & 0 & -2\alpha_{xx} \end{pmatrix}
\qquad
\begin{pmatrix} -\sqrt{3}\,\alpha_{xx} & 0 & 0 \\ 0 & \sqrt{3}\,\alpha_{xx} & 0 \\ 0 & 0 & 0 \end{pmatrix}$$

$T,\ T_g,\ T_2,\ T_2,\ T_{2g}$:
$$\begin{pmatrix} 0 & 0 & 0 \\ 0 & 0 & \alpha_{yz} \\ 0 & \alpha_{yz} & 0 \end{pmatrix}
\qquad
\begin{pmatrix} 0 & 0 & \alpha_{xz} \\ 0 & 0 & 0 \\ \alpha_{xz} & 0 & 0 \end{pmatrix}
\qquad
\begin{pmatrix} 0 & \alpha_{xz} & 0 \\ \alpha_{xz} & 0 & 0 \\ 0 & 0 & 0 \end{pmatrix}$$

experiment with the light incident along the z-axis polarized in the x direction gives a spectrum described by

$$(E_{tx}, E_{ty}, E_{tz}) = \begin{pmatrix} \alpha_{xx} & 0 & 0 \\ 0 & \alpha_{yy} & 0 \\ 0 & 0 & \alpha_{zz} \end{pmatrix} \begin{pmatrix} E_{0x} \\ 0 \\ 0 \end{pmatrix} = (\alpha_{xx}E_{0x}, 0, 0) \quad A_1$$

$$(E_{tx}, E_{ty}, E_{tz}) = \begin{pmatrix} 0 & \alpha_{xy} & 0 \\ \alpha_{xy} & 0 & 0 \\ 0 & 0 & 0 \end{pmatrix} \begin{pmatrix} E_{0x} \\ 0 \\ 0 \end{pmatrix} = (0, \alpha_{xy}E_{0x}, 0) \quad A_2$$

$$(E_{tx}, E_{ty}, E_{tz}) = \begin{pmatrix} 0 & 0 & \alpha_{xz} \\ 0 & 0 & 0 \\ \alpha_{xz} & 0 & 0 \end{pmatrix} \begin{pmatrix} E_{0x} \\ 0 \\ 0 \end{pmatrix} = (0, 0, \alpha_{xz}E_{0x}) \quad B_1$$

$$(E_{tx}, E_{ty}, E_{tz}) = \begin{pmatrix} 0 & 0 & 0 \\ 0 & 0 & \alpha_{yz} \\ 0 & \alpha_{zy} & 0 \end{pmatrix} \begin{pmatrix} E_{0x} \\ 0 \\ 0 \end{pmatrix} = (0, 0, 0) \qquad B_2$$

Thus in this experiment, the Raman scattered beam polarized in the x direction will show a transition associated with the A_1 normal mode of vibration, while the Raman spectra polarized in the y and z directions will show transitions associated with the A_2 and B_1 normal modes of vibration, respectively. In order to see the B_2 phonon mode in the Raman spectrum, a y or z polarization direction of the incident beam must be used. To summarize, the possible polarization combinations for observing each of the normal modes of vibration in this example are

A_1	A_2	B_1	B_2
E_{ox}, E_{tx}	E_{ox}, E_{ty}	E_{ox}, E_{tz}	E_{oz}, E_{ty}
E_{oy}, E_{ty}	E_{oy}, E_{tx}	E_{oz}, E_{tx}	E_{oy}, E_{tz}
E_{oz}, E_{tz}			

Raman scattering is not a resonant process like infrared absorption so it is not restricted to phonon at the center of the Brillouin zone. However, energy and momentum still have to be conserved, so for Stokes scattering

$$\vec{k}_{\text{Stokes}} = \vec{k}_i - \vec{q}_R, \quad \omega_{\text{Stokes}} = \omega_i - \omega_R$$

and for anti-Stokes scattering

$$\vec{k}_{\text{anti-Stokes}} = \vec{k}_i + \vec{q}_R, \quad \omega_{\text{anti-Stokes}} = \omega_i + \omega_R.$$

Combining these expressions results in

$$c^2 q_R^2 = n_i^2 \omega_i^2 + n_S^2 (\omega_i - \omega)^2 - 2 n_i n_S \omega_i (\omega_i - \omega) \cos \varphi, \qquad (7.28)$$

where φ is the angle between the incident and scattered light beams. This expression relates the frequencies and wave vectors of the Raman active phonons observable at different scattering angles in a Raman spectrum.

The experimental technique described above is similar to the use of polarized two-photon spectroscopy to identify the symmetry properties of excited electronic states as described in Sect. 6.5. The quantum mechanical description of Raman scattering involves second-order perturbation theory utilizing the electric dipole interaction Hamiltonian, the electron–phonon interaction Hamiltonian given in (7.14), and the electron and phonon wave functions for the system in Fermi's Golden Rule for transition rates [7]. This leads to the same Raman transition selection rules as found above from group theory considerations, but allows the strength of the Raman scattering cross section to be determined.

Raman spectroscopy is useful in studying the effects of external perturbations on the vibrational modes of a material. The application of an electric field, magnetic field, strain field, or other external force with specific directional properties alters the symmetry of the crystal and thus changes the Raman tensor. The symmetry of the external perturbation can be combined with the symmetry of the unperturbed Raman tensor to determine new selection rules. Another method of treating these types of perturbations is through the use of subgroups and compatibility relations as discussed in Chap. 2. In the example of Raman scattering from a crystal with C_{2v} symmetry discussed above, the application of an electric field in the xy plane will destroy the two σ_v symmetry planes and change the point group to C_2. The compatibility relations between C_{2v} and its C_2 subgroup show that the A_1 and A_2 irreducible representations in the former transform as the A irreducible representation in the latter while the B_1 and B_2 irreducible representations in C_{2v} transform as the B_2 irreducible representation in C_2. The Raman spectra for the polarization conditions E_{ox},E_{tx}; E_{oy},E_{ty}; E_{oz},E_{tz}; E_{ox},E_{ty}; and E_{oy},E_{tx} will contain transitions associated with the A vibrational mode while the spectra with the associated with the B vibrational mode. In come cases an external perturbation can cause a Raman inactive mode to become active.

7.5 Jahn–Teller Effect

Up to this point it has been assumed that the wave functions for the system of interest can be factored into two parts, one describing the electronic state of the system and the other describing the vibrational part of the system. This is known as the *Born–Oppenheimer* approximation and the individual electronic and vibrational wave functions are eigenfunctions of two different Schrödinger equations. This is based on the fact that the motion of the atomic nuclei is much slower than the motion of the electrons. Since the nuclear motion depends on the electronic charge distribution, these equations are coupled. An energy eigenvalue of the electronic Schrödinger equation acts as a potential in the vibrational Schrödinger equation. The eigenvalues of the vibrational Schrödinger equation represent the total energy of the system in specific electronic and vibrational states.

The dependence of the electronic eigenvalue $E_\alpha(X)$ on the positions of the nuclei X can be expressed in terms of an expansion about the equilibrium positions $u_{iv} = (X - X(0))_{iv}$

$$E_\alpha(X) = E_\alpha(X(0)) + \sum_{iv} \left(\frac{\partial E_\alpha}{\partial u_{iv}}\right)_{u=0} u_{iv} + \cdots, \qquad (7.29)$$

where higher order and anharmonic terms have been ignored. The first term on the left represents the potential with the nuclei in their equilibrium positions and the second term represents the change in the potential when the nuclei move away from their equilibrium position. Transforming to normal mode coordinates q as done in Chap. 5 shows that the first term in (7.29) simply adds a constant in the vibrational Schrödinger equation [8] while the second term becomes

$$E_\alpha(X) \approx \sum_{q_i} \left(\frac{\partial E_\alpha}{\partial q_i}\right)_0 q_i. \qquad (7.30)$$

If this term is identically equal to zero the system is in equilibrium.

Taking the derivative with respect to q_i on both sides of the electronic Schrödinger equation and multiplying from the right by the electronic eigenfunction ψ_{el} gives

$$\left(\frac{\partial E_\alpha}{\partial q_i}\right)_0 = \langle \psi_{el} | \frac{\partial H_{el}}{\partial q_i} | \psi_{el} \rangle. \qquad (7.31)$$

This can be rewritten in terms of the irreducible representations of the point group of the system as has been done previously, for example (7.19). The electronic wave functions transform according to irreducible representations Γ_{el}. (For complex representations, one of the two will be the complex conjugate of the other.) The Hamiltonian always transforms according to the totally symmetric A_{1g} representation while the derivative with respect to the normal coordinate transforms according to the same irreducible representation as q_i, Γ_q. Thus the matrix element on the right-hand side of (7.31) equals zero unless

$$\Gamma_{el} x \Gamma_q x \Gamma_{el} \supset A_{1g}. \qquad (7.32)$$

Thus for (7.30) to equal zero so the system is in equilibrium, (7.32) requires that

$$\Gamma_q \not\subset |\Gamma_{el}|^2. \tag{7.33}$$

This follows from (7.32) using the fact that the totally symmetric representation will only appear in the reduction of the product of an irreducible representation with itself.

If the electronic state is nondegenerate, its product transforms as the totally symmetric representation. Under these conditions the coupling to an asymmetric normal vibrational mode will fulfill condition (7.33) and the system remains in equilibrium with the same symmetry as it had without the electron–phonon coupling. If a nondegenerate electronic state couples to a totally symmetric normal vibrational mode the criteria in (7.33) is not met so the electron–phonon interaction does cause a shift from the equilibrium condition. However, since both the electronic and vibrational modes are totally symmetric, no change in system symmetry occurs in the coupled equilibrium condition. The same will be true for a degenerate normal vibrational mode coupled to a degenerate electronic state. The interesting situation occurs when an asymmetric normal vibrational mode is coupled to a degenerate electronic state. For some cases of this type the condition in (7.33) is not met and the electron–phonon coupling forces the system to undergo a change in symmetry to reach a new equilibrium position. This distortion is called the *Jahn–Teller effect* and essentially represents a breakdown of the Born–Oppenheimer approximation.

It was shown by Jahn and Teller [9] that the reduction of the square of each degenerate irreducible representation of all of the symmetry point groups contains at least one irreducible representation for an asymmetric vibrational normal mode. Electron–phonon coupling involving this mode will cause the system to transition to a new equilibrium symmetry. The Schrödinger equation for the coupled electron–phonon system with degenerate electronic states can have multivalued solutions.

As an example, consider an octahedral with the 15 normal modes of vibration shown in Fig. 7.2. Since the electronic states in (7.31) have same parity, the normal vibrational mode must have even parity for this matrix element to be nonzero. One of these is a nondegenerate A_{1g} mode that shifts the energy level of the system but does not change the symmetry and thus does not cause a splitting of the energy level. There are two types of degenerate even parity vibrational modes for a complex with O_h symmetry, a doubly degenerate E_g mode and a triply degenerate T_{2g} mode. As shown in Fig. 7.2, these change the symmetry of the system and therefore can cause a splitting of the energy level. As an example, consider an electronic state that transforms according to the E_g irreducible representation of O_h. In this point group $E_g \times E_g = A_{1g} + A_{2g} + E_g$ so the E_g vibrational mode shown in Fig. 7.2 will cause a Jahn–Teller effect when coupled with an electronic state transforming as the E_g irreducible representation. For an electronic state transforming as T_{2g} in O_h symmetry, the reduction of its direct product with itself is $T_{2g} \times T_{2g} = A_{1g} + E_g + T_{1g} + T_{2g}$. Thus vibrational modes shown in Fig. 7.2 transforming either as E_g or T_{2g} can couple with a T_{2g} electronic state to produce a Jahn–Teller effect.

Titanium doped sapphire, Al_2O_3:Ti^{3+}, is an important tunable solid state laser material [2]. The free ion energy level of the single d electron associated with trivalent titanium is split by the primarily octahedral crystal field at the site of the aluminum ion in the sapphire lattice to give a ground state electronic level transforming as T_{2g} and an excited state level transforming as E_g as shown in Fig. 4.3. The E_g excited electronic state couples with the E_g vibrational mode of the TiO_6 octahedral complex to create a Jahn–Teller effect. This causes the electronic potential well of the excited state to have two energy minima both of which are off-set in position coordinates from the energy minimum of the ground state potential well. The absorption shows two unresolved broad bands due to transitions from the ground state to the two Jahn–Teller split components of the excited state. The Jahn–Teller splitting of the excited state plays an important role in determining the spectroscopic properties of titanium-doped sapphire. The T_{2g} ground state also undergoes a Jahn–Teller distortion but it is much smaller and not as important in determining the optical properties of the material.

Figure 7.7 shows the energy levels involved in the situation described in the previous paragraph. When the electronic and vibrational states of the system can be treated separately, the electronic states have the shape of a parabola with respect to a lattice coordinate q due to the modulation of q by a lattice vibration. The different vibrational states are designated as horizontal lines in each electronic potential well. A typical situation of this type is depicted in Fig. 7.7a Note that having different electron–phonon interactions in the ground and excited states can cause the minima of the two electronic potentials to be located at different values of q. The vertical arrow shows the absorption transition from the ground to the excited state. Figure 7.7b shows the same situation when the excited state is a strongly coupled vibrational-electronic (vibronic) state that exhibits Jahn–Teller splitting. Assuming

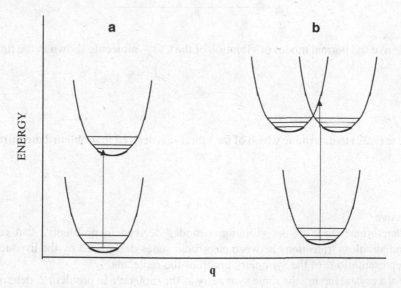

Fig. 7.7 Configuration-coordinate diagrams for a system in the Born–Oppenheimer approximation (a) and with the Jahn–Teller effect (b)

that both the electronic part of the coupled state and the vibrational part transform as the E_g irreducible representation in O_h symmetry as discussed above, the vibronic state is described by two parabolic potential wells with their minima occurring at different values of q. The absorption transition can occur to both of the excited potential well. The Jahn–Teller energy is the difference in energy between the bottom of one of the potential wells and the energy at which the two cross. The Jahn–Teller effect is especially important for transition metal ion with unshielded d electrons.

7.6 Problems

1. Derive the lattice vibrations for a two-dimensional crystal with C_{4v}^1 symmetry as shown in the figure.

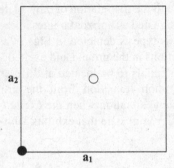

2. Derive the normal modes of vibration of the CO_3^{2-} molecule shown in the figure.

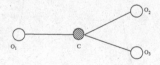

3. Use (7.22) to determine which of the vibrations derived in problem 2 are infrared active.

4. Determine which of the vibrational modes derived in problem 2 can cause radiationless transitions between electronic states designated by the irreducible representations of the symmetry group of the molecule.

5. For a crystal having the same symmetry as the molecule in problem 2, determine which vibrational modes will be active in a Raman scattering experiment with light incident along the z-axis polarized in the y direction.

References

1. F.A. Cotton, *Chemical Applications of Group Theory* (Wiley, New York, 1963)
2. R.C. Powell, *Physics of Solid State Laser Materials* (Springer, New York, 1998)
3. E.B. Wilson, J.C. Decius, P.C. Cross, *Molecular Vibrations* (McGraw-Hill, New York, 1955)
4. B. DiBartolo, Optical Interactions in Solids (Wiley, New York, 1968)
5. B. Dibartolo, R.C. Powell, *Phonons and Resonances in Solids* (Wiley, New York, 1976)
6. Q. Kim, R.C. Powell, M. Mostoller, T.M. Wilson, Phys. Rev. B **12**, 5627 (1975); Q. Kim, R.C. Powell, T.M. Wilson, Solid State Commun. **14**, 541 (1974); Q. Kim, R.C. Powell, T.M. Wilson, J. Phys. Chem. Solids **36**, 61 (1975)
7. R. Loudon, Proc. Roy. Soc. A **275**, 218 (1963)
8. R.S. Knox, A. Gold, *Symmetry in the Solid State* (Benjamin, New York, 1964)
9. H.A. Jahn, E. Teller, Proc. Roy. Soc. **A161**, 220 (1937)

Chapter 8
Symmetry and Electron Energy Levels

Electrons in a crystalline solid interact with the atoms on the lattice and with other electrons. Some of the electrons remain part of their parent atom. However, others take part in bonding the atoms together to form the crystal lattice and others may become quasi-free electrons that can move about in the solid. Chapter 4 discussed the importance of symmetry in determining the properties of electron energy levels of an isolated ion in a crystal field. In the examples given, the d and f electron orbitals were assumed to have no overlap with electrons on neighboring atoms. The bonding was purely ionic in nature and any covalency was treated as a minor perturbation on the results. In this chapter the importance of symmetry in describing the properties of chemical bonding is discussed and then the use of symmetry in determining the properties of energy bands of quasi-free electrons is described. The dispersion relations for electronic energy bands are similar to the phonon dispersion properties described in Chap. 7.

8.1   Symmetry and Molecular Bonds

The first step in treating molecular bonding is to determine the combinations of electron orbitals on an atom that can be used to construct hybrid orbitals that can form bonds with neighboring atoms. The most common types of electron orbitals used to create these combinations are the s ($l=0$), p ($l=1$), and d ($l=2$) orbitals described in Chap. 4. As discussed previously, the wave functions for these orbitals can be factored into a radial part and an angular part each of which are normalized separately. The spherical coordinate system depicting this is shown in Fig. 6.5 with the radial extent of the orbital measured as the distance from the origin and the two angular directions giving the spatial orientation as shown. Symmetry considerations have no effect on the on the radial part of the wave function so only the angular factor must be considered. The relationships between the Cartesian and spherical coordinates in Fig. 6.5 are

R.C. Powell, *Symmetry, Group Theory, and the Physical Properties of Crystals*,
Lecture Notes in Physics 824, DOI 10.1007/978-1-4419-7598-0_8,
© Springer Science+Business Media, LLC 2010

$$x = r \sin \theta \cos \varphi$$
$$y = r \sin \theta \sin \varphi \qquad (8.1)$$
$$z = r \cos \theta$$

As discussed in Chap. 4, the angular parts of the wave functions of the electron orbitals on an atom can be expressed in terms of the spherical harmonic functions listed in Table 4.1. The expressions for the angular parts of the electron wave functions of interest here are shown explicitly in Table 8.1.

Figure 8.1 shows some examples of the spatial distributions of s, p, and d wave functions. The plus and minus signs indicate the sign of the wave function at that point in space as determined by the mathematical expressions in Table 8.1. The s wave function is spherical symmetric and positive at all points in space. There are three p wave functions shaped as figure eights oriented along the three coordinate axes. The p_y orbital is shown in Fig. 8.1. As noted in the figure, it has a positive sign for the part in the $+y$ direction and a negative sign for the part in

Table 8.1 Angular factors of s, p, and d electron orbitals

Orbital	$\psi_l^{m_l}(\theta, \varphi)$
s	$1/(2\sqrt{\pi})$
p_x	$\left[\sqrt{3}/(2\sqrt{\pi})\right] \sin \theta \cos \varphi$
p_y	$\left[\sqrt{3}/(2\sqrt{\pi})\right] \sin \theta \sin \varphi$
p_z	$\left[\sqrt{3}/(2\sqrt{\pi})\right] \cos \theta$
d_{z^2}	$\left[\sqrt{5}/(4\sqrt{\pi})\right] (3 \cos^2 \theta - 1)$
d_{xz}	$\left[\sqrt{15}/(2\sqrt{\pi})\right] (\sin \theta \cos \theta \cos \varphi)$
d_{yz}	$\left[\sqrt{15}/(2\sqrt{\pi})\right] (\sin \theta \cos \theta \sin \varphi)$
$d_{x^2-y^2}$	$\left[\sqrt{15}/(4\sqrt{\pi})\right] (\sin^2 \theta \cos 2\varphi)$
d_{xy}	$\left[\sqrt{15}/(4\sqrt{\pi})\right] (\sin^2 \theta \sin 2\varphi)$

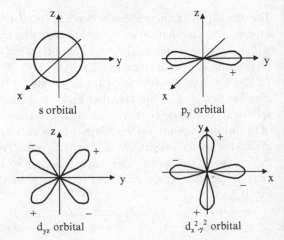

Fig. 8.1 Examples of the spatial distributions of some of the s, p, and d single electron orbitals

the $-y$ direction. The other two orbitals p_x and p_z have similar orientations and signs with respect to their axes. There are five d orbitals three of which are in the xy, xz, and yz planes. These have the shape of four-leaf clovers with their lobes oriented between the axes. The example in the figure is the d_{yz} orbital which is positive in the first and third quadrants and negative in the second and fourth quadrants. The other two orbitals are each similar to this example with the appropriate orientations in their planes. The other two orbitals are designated d_{z^2} and $d_{x^2-y^2}$. They also have a four-leaf clover shape but the lobes are oriented along the coordinate axes directions. The example shown in Fig. 8.1 is the $d_{x^2-y^2}$ orbital which lies in the xy plane with lobes along the x and y directions. The signs of both lobes along the x-axis are negative while the lobes along the y-axis have positive signs. The d_{z^2} orbital has a similar shape in the xz plane except that the positive lobes along the z-axis are larger in magnitude than the negative lobes oriented along the x-axis.

In Chap. 4 the fact that spherical harmonic functions can transform as basis functions for irreducible representations of symmetry point groups was discussed. The character tables in Chap. 2 designate how coordinates x, y, and z and products such as z^2, yz, and x^2+y^2 transform for each of the 32 symmetry groups. (Note that x^2+y^2 transforms in the same was as z^2.) These functions appear as subscripts on the orbital designations in Table 8.1 and the orbital transforms according to the same representation as its subscript. An s orbital always transforms at the totally symmetric representation. For a molecule having C_{3v} symmetry, Table 2.17 shows that the s, p_z, and d_{z^2} orbitals all transforms as the A_1 representation. The p_x and p_y orbitals transform together as a basis for the doubly degenerate E representation. The d_{xy} and $d_{x^2-y^2}$ orbitals transform together as the E irreducible representation and the d_{yz} and d_{xz} orbitals form another set of functions transforming as E. Similar assignments can be made for any of the symmetry point groups using the character tables given in Chap. 2.

The symmetry properties of multielectron atoms are constructed from the properties of the single electron orbitals discussed above. The combinations of these orbitals form *hybrid orbitals* that reflect the symmetry properties of the molecule and its bonding characteristics. As an example, consider a molecule AB_4 with tetrahedral T_d symmetry as shown in Fig. 8.2. For this case there must

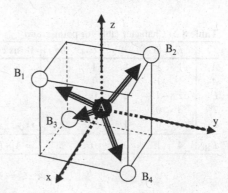

Fig. 8.2 Tetrahedral molecule AB_4

be four hybrid orbitals on atom A constructed so their wave functions point to the four B atoms [1]. The four large vectors in Fig. 8.2 represent the hybrid orbitals. These can be labeled r_n where n runs from 1 to 4 for the different B_n atoms. The set of four r_n hybrid orbitals transform as a basis for a reducible representation in T_d symmetry. The characters for this representation can be found by applying the symmetry operations of the group to the four vectors. Each symmetry operation is expressed as a 4×4 matrix with the 16 elements being 1 if the vector labeling that row transforms into the vector labeling the column. Only the vectors unchanged by the operation will have 1s along the diagonal of the matrix and thus contribute to the character. There are five classes of symmetry operations for T_d symmetry as shown in Table 2.30. The identity operation leaves all the vectors invariant and thus has a character of 4. The rotation of 180° about the z-axis interchanges r_1 with r_2 and r_3 with r_4 and thus has a character of 0. The threefold rotation axes run along the diagonals of the cube shown in Fig. 8.2 and therefore each of them will contain one of the vectors. That one vector will remain unchanged while the other three will transform into each other. This gives $\chi(C_3)=1$. The diagonal mirror planes containing the face diagonals of the cube will leave two of the vectors unchanged and interchange the other two giving a character of $\chi(\sigma_d)=2$. Finely, the improper rotations involving 90° rotations about the x, y, or z axis followed by reflections in planes perpendicular to the rotation axes leave none of the vectors invariant which gives $\chi(S_4)=0$. These results are summarized in the character table for T_d symmetry repeated here as Table 8.2.

As seen in Table 8.2, the hybrid orbitals shown in Fig. 8.2 transform as a reducible representation in the T_d symmetry group of the molecule. The bottom line of the table shows the characters of this representation and (2.10) (or inspection) can be used to reduce this in terms of the irreducible representations A_1 and T_2. From the discussion above and the basis components shown in the final column of the table, the presence of the A_1 irreducible representation in Γ_{HO} shows that a single electron s orbital makes up part of the hybrid orbitals. The presence of the triply degenerate T_2 irreducible representation in the reduction of Γ_{HO} shows that either the set of three p_x, p_y, and p_z orbitals or the set of three d_{xy}, d_{xz}, and d_{yz} orbitals are part of the hybrid orbitals. These two possibilities for the hybrid orbitals are designated sp^3 and sd^3. In general, the hybrid orbitals in an AB$_4$ molecule will

Table 8.2 Character table for point group T_d

T_d	E	$8C_3$	$3C_2$	$6S_4$	$6\sigma_d$	Basis components	
A_1	1	1	1	1	1		$x^2+y^2+z^2$
A_2	1	1	1	−1	−1		
E	2	−1	2	0	0		$(2z^2-x^2-y^2, x^2-y^2)$
T_1	3	0	−1	1	−1	(R_x, R_y, R_z)	
T_2	3	0	−1	−1	1	(x,y,z)	(xz,yz,xy)
Γ_{HO}	4	1	0	0	2	$= A_1+T_2 =$ s + (p$_x$, p$_y$, p$_z$) or s + (d$_{xy}$,d$_{xz}$,d$_{yz}$)	

be a mixture of these two types if the single electron orbitals are all in the same energy shell (i.e., the same n quantum number). If the orbitals come from different energy shells, one type of hybrid might be energetically favored another type.

In the example above, the hybrid orbitals form σ bonds; in other words the wave functions do not have zero amplitudes and change signs on a surface that contains the bond axes. Wave functions that have one nodal surface containing the bond axis are called π bonds. For $\pi-\pi$ bonding between two atoms in a molecule, each atom must contribute an orbital with its nodal plan aligned parallel to the nodal plane of the orbital on the other atom. There are two orthogonal conditions for this to occur. As an example consider a planar AB_3 molecule with D_{3h} symmetry as shown in Fig. 8.3. Each B molecule will have two possible orthogonal π orbitals and the A molecule will have a set of π orbitals equal to the total number of π orbitals on the B atoms and aligned in the same directions [1]. The set of orbitals on B and the set on A both transform according to the same reducible representation of the symmetry group of the molecule. The characters of this representation are found in the same way as done in the previous example except that the π bonds are represented by two orthogonal vectors on each of the B atoms as shown in the figure. The four atoms are in the xy plane with the z-axis coming out of the plane at the center of the A atom. The vectors labeled B_{n1} are in the plane of the molecule while the dots labeled B_{n2} indicate vectors coming out of the page parallel to the z-axis. This set of six vectors B_{ij} are the functions used to derive the characters of the hybrid orbital representation.

The character table for the D_{3h} symmetry group was given in Chap. 2 and is repeated here for convenience in Table 8.3. None of the symmetry operations of this group will transform one of the vectors perpendicular to the plane into one of the vectors in the plane. Thus these two sets of three vectors can be treated separately. The three vectors in the plane transform according to a representation designated Γ_1 while the three vectors perpendicular to the plane transform according to a representation labeled Γ_2. First consider the three vectors in the plane. The identity operation leaves all three of them invariant and thus as a character $\chi_1(E)=3$. The threefold rotation about the z-axis interchanges all three of these vectors so $\chi_1(C_3)=0$. The twofold rotations about the dashed lines shown in Fig. 8.3 each reverses the direction of one vector and interchanges the directions of the others giving a

Table 8.3 Character table for point group D_{3h}

D_{3h}	E	$2C_3$	$3C_2$	σ_h	$2S_3$	$3\sigma_v$	Basis components	
A'_1	1	1	1	1	1	1		x^2+y^2,z^2
A'_2	1	1	−1	1	1	−1	R_z	
E'	2	−1	0	2	−1	0	(x,y)	(x^2-y^2,xy)
A''_1	1	1	1	−1	−1	−1		
A''_2	1	1	−1	−1	−1	1	z	
E''	2	−1	0	−2	1	0		(R_x,R_y) (xz,yz)
Γ_1	3	0	−1	3	0	−1	$= A'_2 + E' = (p_x,p_y) + (d_{xy},d_{x^2-y^2})$	
Γ_2	3	0	−1	−3	0	1	$= A''_2 + E'' = p_z + (d_{xz},d_{yz})$	
Γ_{HO}	6	0	−2	0	0	0		

Fig. 8.3 AB$_3$ molecule with D_{3h} symmetry

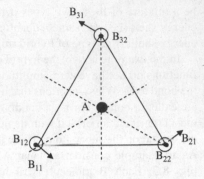

character of -1. The horizontal mirror plane leaves all three vectors invariant giving a character of 3. None of the vectors are invariant under the threefold improper rotation giving it a character of 0. Finally the vertical mirror planes containing the horizontal lines shown in the figure each reverses the direction of one vector and interchange the other two. This leads to a character of -1. These results are summarized in Table 8.3. Similar considerations of the transformation properties of the three vectors perpendicular to the plane of the molecule give the characters of the Γ_2 representation shown in the table.

The total representation for the hybrid orbitals is the sum of the horizontal and perpendicular orientations. As shown in the last column in Table 8.3, the reducible representations for both the horizontal and perpendicular orientations can be reduced in terms of one one-dimensional and one two-dimensional irreducible representations. These four irreducible representations then appear in the reduction of the reducible representation of the total hybrid orbital representation. The symmetry property of single electron orbitals shown in Table 8.1 along with the basis functions shown in the final column of Table 8.3 shows that for atom A to form a π bond with each of the B atoms oriented perpendicular to the plane it must form a hybrid bond constructed from its p$_z$, d$_{xz}$, and d$_{yz}$ single electron orbitals. It is not possible for atom A to form three equivalent π bonds oriented in the plane of the molecule since none of the s, p, and d orbitals transform as the A_2' irreducible representation. It still may be possible to form two π bonds with this orientation using the degenerate sets of p and d orbitals that transform as the E irreducible representation. In this case the two bonds must be shared among the three B atoms.

The mathematical expressions for a specific type of hybrid orbital can be written as a linear combination of single electron wave functions. This technique is referred to as LCAO, linear combination of atomic orbitals. It is useful in calculating the strengths of chemical bonds. In this approach the ith hybrid orbital Ψ_i written as a linear expansion of single electron atomic orbitals ψ_n

$$\Psi_i = a_{i1}\psi_1 + a_{i2}\psi_2 + a_{i3}\psi_3 + \cdots, \tag{8.2}$$

where the a_{in} are the expansion coefficient which must be evaluated. Since the individual atomic orbitals are orthonormal, for Ψ_i to be normalized

$$a_{i1}^2 + a_{i2}^2 + a_{i3}^2 + \cdots = 1. \tag{8.3}$$

Also, for different Ψ_i to be orthogonal to each other,

$$a_{i1}a_{j1} + a_{i2}a_{j2} + a_{i3}a_{j3} + \cdots = 0. \tag{8.4}$$

A set of hybrid orbitals results in a set of simultaneous equations of the form of (8.2). The values of the expansion coefficients must be consistent with the transformation of one hybrid orbital to an equivalent one under a symmetry operation of the group. These criteria can be used to evaluate the expansion coefficients.

As an example consider a set of sp^2 hybrid orbitals in the xy plane. These can be visualized using Fig. 8.3 ignoring the π orbital vectors and working with Cartesian coordinate axes. The x-axis is taken to point along direction 1 in the figure, the y-axis perpendicular to this pointing between the 1 and 2 directions, and the z-axis coming out of the paper at the central atom. This set of orbitals has D_{3h} symmetry with the character table shown in Table 8.3. Selecting the first orbital of the set to be along the direction of the x-axis gives

$$\Psi_1 = a_{1s}s + a_{1x}p_x + a_{1y}p_y.$$

Since the orbital p_y has no amplitude along the x-axis the value of a_{1y} is equal to 0. The second hybrid orbital of the set will be in the second quadrant of the xy plane where the sign of the p_x wave function is negative. By convention, the expansion coefficients are taken to be positive and the sign of the term insures a positive contribution of the single electron orbital. Thus,

$$\Psi_2 = a_{2s}s - a_{2x}p_x + a_{2y}p_y.$$

The final hybrid orbital of the set will be in the third quadrant where the signs of both the p_x and p_y wave functions are negative so

$$\Psi_3 = a_{3s}s - a_{3x}p_x - a_{3y}p_y.$$

The symmetry of the system can now be used to evaluate the eight remaining expansion coefficients. For example, the 60° rotation in the clockwise direction about the z-axis takes Ψ_2 into Ψ_1. Applying this to the equations above gives

$$C_3\Psi_2 = \Psi_1$$

or

$$C_3\left[a_{2s}s - a_{2x}p_x + a_{2y}p_y\right] = \left[a_{2s}C_3s - a_{2x}C_3p_x + a_{2y}C_3p_y\right] = a_{1s}s + a_{1x}p_x.$$

Using the facts that s is invariant under all operations of the group while p_x transforms like a unit vector in the x direction and p_y transforms like a unit vector in the y direction gives

$$a_{2s}s + \left(\frac{1}{2}a_{2x} + \frac{\sqrt{3}}{2}a_{2y}\right)p_x + \left(\frac{\sqrt{3}}{2}a_{2x} - \frac{1}{2}a_{2y}\right)p_y = a_{1s}s + a_{1x}p_x.$$

For the symmetry condition to hold, the coefficients of each orbital on both sides of the equation must be equal:

$$a_{2s} = a_{1s}, \quad a_{2y} = \sqrt{3}a_{2x}, \quad a_{1x} = 2a_{2x}.$$

Next consider the σ_{xz} reflection plane that will transform Ψ_2 into Ψ_3 and leave Ψ_1 invariant. This operation on the expressions for the LCAO wave functions gives

$$\sigma_{xz}\Psi_3 = \Psi_2$$

or

$$\sigma_{xz}\left[a_{3s}s - a_{3x}p_x - a_{3y}p_y\right] = \left[a_{3s}s - a_{3x}p_x + a_{3y}p_y\right] = a_{2s}s - a_{2x}p_x + a_{2y}p_y,$$

where the fact that the s and p_x orbitals are invariant under this reflection and p_y is transformed into $-p_y$ has been used. For this equation to hold

$$a_{3s} = a_{2s}, \quad a_{3x} = a_{2x}, \quad a_{3y} = a_{2y}.$$

Next the conditions of normalization and orthogonality can be used to provide two more equations to complete the set of eight equations to solve for the eight unknowns. The results give the final expressions for the three hybrid orbitals to be

$$\Psi_1 = \frac{1}{\sqrt{3}}s + \frac{2}{\sqrt{6}}p_x$$

$$\Psi_2 = \frac{1}{\sqrt{3}}s - \frac{1}{\sqrt{6}}p_x + \frac{1}{\sqrt{2}}p_y. \tag{8.5}$$

$$\Psi_3 = \frac{1}{\sqrt{3}}s - \frac{1}{\sqrt{6}}p_x - \frac{1}{\sqrt{2}}p_y$$

This LCAO method can be extended to form sets of orbitals extending over the entire molecule called *molecular orbitals*. Using similar symmetry techniques described above, explicit expressions for the molecular orbital in terms of the atomic orbitals can be obtained and the energy levels found from the secular determinant. This provides expressions for the energy levels of the molecule and thus determines which ones are stable bonding orbitals and which are not. The general form of a molecular orbital is

$$\Psi = \psi_A(\Gamma) + \sum_i a_i \psi_{Bi}, \qquad (8.6)$$

where the first term on the right is a wave function of the central atom that transforms according to the Γ irreducible representation of the symmetry group of the molecule while the second term on the right is a hybrid orbital constructed from a linear combination of the orbitals of the ligands. This hybrid orbital must transform as the same irreducible representation as the first term so that the molecular orbital transforms as Γ.

As an example, consider an octahedral molecule made up of a central atom and six ligands as shown in Figs. 2.3, 4.1, and 7.1. The normal modes of vibration of this type of molecule with O_h symmetry were derived in Sect. 7.1. The coordinate system for constructing the molecular orbitals is shown in Fig. 8.4. Note that the numbering of the atoms and the orientations of the coordinates on each atom is different from Fig. 7.1. Figure 8.4 follows the convention for numbering the ligands and the directions of the coordinates commonly used for the construction of molecular orbitals [2]. The z-axis always point from the ligand toward the central ion. This results in the σ bonds being described by the z coordinate and the π bonds by the x and y coordinates. Also, the central atom has a right-handed coordinate system while each of the ligands has left-handed coordinate systems.

First consider σ bonds which are symmetric about the line joining the ligand with the central atom. The character table for O_h symmetry given in Table 2.32 shows that the central ion will have s orbitals transforming as a_{1g}, p_n orbitals transforming as t_{1u}, and d_n orbitals transforming as the e_g and t_{2g} irreducible representations. There will be six σ bonds constructed from linear combinations of the s, p_z, and d_{z^2} single electron orbitals of the ligands. They will be mutually orthogonal and each will

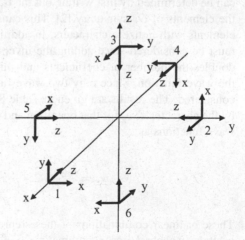

Fig. 8.4 Coordinates for constructing molecular orbitals in a molecule with O_h symmetry

transform as one of the irreducible representations of the O_h point group. The procedure for finding the characters of the representation of the σ bonds is exactly the same as described in Sect. 7.1 for finding the characters of the radial vibrational modes. The character for a specific symmetry element equals the sum the number of ligands left in the same position after the symmetry element is applied to the molecule. This is because these contribute to the diagonal elements of the transformation matrix for that symmetry operation. Table 7.1 shows the result of this analysis so it will not be repeated here. As demonstrated, the reducible representation obtained in this way can be reduced in terms of the a_{1g}, e_g, and t_{1u} irreducible representation of the O_h symmetry group. Matching the irreducible representations of the central atom orbitals with those of the ligands shows that the hybrid bond will include an s orbital, three p orbitals, and two d orbitals, the last of these transforming as the degenerate e_g irreducible representation.

Using the procedure outlined above, the exact form of the molecular orbitals can be obtained. Since the σ bonds only involve the z coordinates, the normalized wave function transforming as a_{1g} is

$$\psi_s = \frac{1}{\sqrt{6}} (z_1 + z_2 + z_3 + z_4 + z_5 + z_6). \tag{8.7}$$

For the three p orbitals transforming as a basis for the t_{1u} representation, instead of going through the formal procedure to obtain the wave functions, it can be seen by inspection that the maximum interaction between the central ion and ligand orbitals is obtained from wave functions of the form

$$\psi_{p_z} = \frac{1}{\sqrt{2}} (z_3 - z_6), \quad \psi_{p_y} = \frac{1}{\sqrt{2}} (z_2 - z_5), \quad \psi_{p_x} = \frac{1}{\sqrt{2}} (z_1 - z_4). \tag{8.8}$$

This leaves the two d wave functions transforming as e_g to be determined. These can be determined by first writing out the transformation table for the z-vectors for the elements of O_h symmetry [2]. This can be simplified by considering only the elements with nonzero characters. In addition, only the elements in O symmetry must be considered since adding the inversion operation and its products simply doubles the number of coefficients and this is lost in the final normalization of the wave function. Since only two wave functions are needed, only z_1 and z_2 are considered. The results are given in Table 8.4. Multiplying each entry in the table by the character of e_g for that operation and combining the like terms gives the two wave functions as

$$\psi_1 = 2z_1 - z_2 - z_3 + 2z_4 - z_5 - z_6,$$
$$\psi_2 = -z_1 + 2z_2 - z_3 - z_4 + 2z_5 - z_6. \tag{8.9}$$

These or linear combinations of these represent the wave functions for the two d orbitals. In general, linear combinations transforming like the d_{x^2} and $d_{x^2-y^2}$ are used.

Table 8.4 Transformation table for z_1 and z_2 for specific elements in O_h symmetry

	E	$C_3(1)$	$C_3(2)$	$C_3(3)$	$C_3(4)$	$C_3(5)$	$C_3(6)$	$C_3(7)$	$C_3(8)$	$C_2(1)$	$C_2(2)$	$C_2(3)$
z_1	z_1	z_3	z_2	z_6	z_2	z_3	z_5	z_5	z_6	z_1	z_4	z_4
z_2	z_2	z_1	z_3	z_6	z_3	z_4	z_3	z_4	z_6	z_2	z_5	z_5

The appropriate linear combination can be found by requiring it to transform like these two d wave functions. For example, since

$$C_4 \begin{pmatrix} d_{z^2} \\ d_{x^2-y^2} \end{pmatrix} = \begin{pmatrix} 1 & 0 \\ 0 & -1 \end{pmatrix} \begin{pmatrix} d_{z^2} \\ d_{x^2-y^2} \end{pmatrix},$$

it follows that

$$C_4(a\psi_1 + b\psi_2) = \pm(a\psi_1 + b\psi_2)$$

so $a = \pm b$. Combining the expressions for ψ_1 and ψ_2 found above in this way and normalizing gives the two d wave functions transforming as e_g to be

$$\psi_{d_{z^2}} = \frac{1}{3\sqrt{2}}(-z_1 - z_2 + 2z_3 - z_4 - z_5 + 2z_6)$$

$$\psi_{d_{x^2-y^2}} = \frac{1}{2}(z_1 - z_2 + z_4 - z_5)$$

(8.10)

The next step is to determine the expressions for the π bonds. These use the x and y arrows on each ligand to determine the transformation properties of the bonds. The characters of the reducible representation in O_h symmetry that has these 12 vectors as it basis are $\chi(E)=12$, $\chi(C_3)=0$, $\chi(C_2)=-4$, $\chi(C_4)=0$, $\chi(C_2')=0$, $\chi(i)=0$, $\chi(S_6)=0$, $\chi(\sigma_h)=0$, $\chi(S_4)=0$, and $\chi(\sigma_d)=0$. Using Table 7.1, this representation can be reduced into a t_{1g}, t_{1u}, t_{2g}, and t_{2u} irreducible representations. The ligand orbitals must also transform according to these irreducible representations.

As an example, consider the t_{2g} representation. The d_{xy}, d_{xz}, and d_{yz} orbitals transform according to this irreducible representation. Thus from Figs. 8.1 and 8.4 the vectors representing the π orbitals on the ligands interacting with a d_{yz} orbital on the central atom are shown in Fig. 8.5. From this figure it is possible to write the expression for the π orbital d_{yz} transforming as one of the basis set for the t_{2g} irreducible representation. Doing the same analysis for the d_{xz} and d_{xy} orbitals gives the set

$$\psi_{d_{xy}}(t_{2g}) = (x_1 + y_2 + x_5 + y_4), \quad \psi_{d_{xz}}(t_{2g}) = (x_3 + y_1 + x_4 + y_6),$$

$$\psi_{d_{yz}}(t_{2g}) = (x_2 + y_3 + x_6 + y_5).$$

(8.11)

The expressions for the wave functions transforming as t_{1u}, t_{1g}, and t_{2u} can be found in the same way.

Fig. 8.5 Example of a d_{yz} orbital of the central atom and the vectors representing p bonding on the ligands

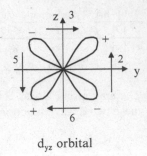

d_{yz} orbital

Now that expressions for the hybrid σ and π ligand orbitals have been determined, (8.6) can be used to determine the molecular orbital. For example, one of the molecular σ bonding orbitals with e_g symmetry is given by

$$\psi_{d_{x^2-y^2}}(e_g) = \alpha d_{x^2-y^2} + \frac{\beta}{2}(z_1 - z_2 + z_4 - z_5).$$

The mixing coefficients α and β determine the amount of time the electron is shared between the central atom and ligand orbital. Solving the secular equations to determine these coefficients leads to two or more solutions. One will have energy less than the orbit on the isolated atoms and is called a bonding orbital while one solution has energy greater than that of the isolated atom and is called an antibonding orbital. The third solution that occurs in some cases is called nonbonding. Figure 8.6 summarizes the molecular orbitals for the octahedral molecule [2]. The lowest three molecular orbitals are bonding, the upper three are antibonding, and the middle one is nonbonding. The a_{1g} orbital of the central ion interacts with the a_{1g} ligand orbital to give both an $a_{1g}(\sigma)$ bonding and an $a_{1g}(\sigma)$ antibonding molecular energy level. The same is true for the e_g orbitals and for the t_{1u} orbitals of the central ion and ligands. The three central ion orbitals d_{xy}, d_{yz}, and d_{zx} of the central ion transforming as t_{2g} do not interact with the ligands and are therefore nonbonding. A similar diagram can be constructed for π bonding.

8.2 Character Tables for Space Groups

In Chap. 2 the character tables for the 32 crystallographic points were given. To deal with delocalized phenomena in solids it is necessary to have a similar character table for the translation group discussed in Chap. 1. The translation operations describing a crystal lattice form a symmetry group that is cyclic. Therefore if T is a translation operation, T^n (where n is an integer) is also a symmetry operation in the group. If there are N_j unit cells in the lattice, the translation group will be a cyclic group of order N_j that has N_j classes. Since all the irreducible representations are one dimensional, there will be N_j irreducible representations for this group [2]. The character table for this type of group is shown in Table 8.5.

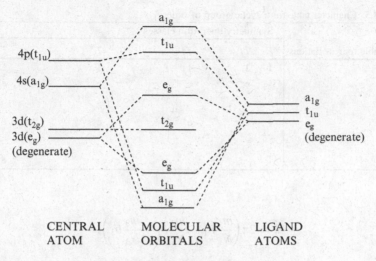

$$4p(t_{1u})$$

$$4s(a_{1g})$$

$$3d(t_{2g})$$
$$3d(e_g)$$
(degenerate)

$$a_{1g}$$
$$t_{1u}$$

$$e_g$$

$$t_{2g}$$

$$a_{1g}$$
$$t_{1u}$$
$$e_g$$
(degenerate)

$$e_g$$
$$t_{1u}$$
$$a_{1g}$$

CENTRAL   MOLECULAR   LIGAND
ATOM   ORBITALS   ATOMS

Fig. 8.6 Molecular orbitals for σ bonding in an octahedral molecule

The trivial representation has 1 for its character for each class of operations as always. The other characters are constructed knowing that $T_j^{N_j} = E$. One way to insure this is through the expression

$$\Gamma_m(T_j) = e^{-2\pi i m/N_j} \quad (m = 0, 1, 2, \ldots, N_j - 1). \tag{8.12}$$

The generic character in Table 8.5 is written as

$$\chi_m\left(T_j^{n_j}\right) = e^{-\left(2\pi i m/N_j\right)n_j} = e^{-ik_j x_j}, \tag{8.13}$$

where the integer n_j has been associated with a translation in the x direction through $x_j = n_j t_j$. In addition, a component of the wave vector has been introduced as

$$k_j = \frac{2\pi}{t_j} \frac{m}{N_j}. \tag{8.14}$$

This shows that irreducible representations of the translation group can be expressed in terms of waves on the periodic lattice.

The irreducible representations for the translation group in three dimensions can be formed as the direct product representation of the three one-dimensional representations [3]. Equations (8.12) and (8.14) then become

$$\Gamma_{m_1 m_2 m_3}\left(\vec{T}\right) = \exp\left\{-2\pi i \left(\frac{n_1 m_1}{N_1} + \frac{n_2 m_2}{N_2} + \frac{n_3 m_3}{N_3}\right)\right\} \tag{8.15}$$

and

Table 8.5 Character table for a cyclic group of order N_j

Irreducible representations	Symmetry operation classes					
	E	T_j	T_j^2	$\cdots$	$T_j^{n_j}$	$\cdots$	$T_j^{N_j-1}$
Γ_0	1	1	1		1		1
Γ_1	1	$e^{-2\pi i/N_j}$	$e^{-4\pi i/N_j}$	$\cdots$	$e^{-2\pi i n_j/N_j}$	$\cdots$	$e^{+2\pi i/N_j}$
$\vdots$	$\vdots$	$\vdots$	$\vdots$	$\vdots$			
Γ_m	1	$e^{-2\pi i m/N_j}$	$e^{-2\pi i m/N_j}$	$\cdots$	$e^{-2\pi i n_j m/N_j}$	$\cdots$	$e^{+2\pi i m/N_j}$
$\vdots$	$\vdots$	$\vdots$	$\vdots$		$\vdots$		$\vdots$
Γ_{N_j-1}	1	$\cdots$	$\cdots$		$\cdots$		$e^{+2\pi i/N_j}$

$$\vec{k} = 2\pi\left(\frac{m_1}{N_1}\vec{b}_1 + \frac{m_2}{N_2}\vec{b}_1 + \frac{m_3}{N_3}\vec{b}_3\right). \tag{8.16}$$

The reciprocal space vectors and real space lattice vectors obey the relationship given in (1.3), $\vec{b}_i \cdot \vec{t}_j = \delta_{ij}$. The expression in (8.13) for the character of a symmetry operation can be written in three dimensions as

$$\chi_{\vec{k}}\left(\vec{R}\right) = e^{-\vec{k}\cdot\vec{R}}. \tag{8.17}$$

The primitive translations in real space $\vec{t}$ and reciprocal space $\vec{b}$ are related by the expressions given in (1.4) and the latter are used to define the Brillouin zone as discussed in Chap. I. Each Brillouin zone has one $\vec{k}$ vector for each irreducible representation of the translation group. A translation in reciprocal space is represented by

$$\vec{K}_j = 2\pi\left(j_1\vec{b}_1 + j_2\vec{b}_2 + j_3\vec{b}_3\right), \tag{8.18}$$

where the j_i are integers. This operation takes $\vec{k} \rightarrow \vec{k} + \vec{K}_j$ and leaves the character given in (8.17) invariant.

8.3 Electron Energy Bands

The physical properties of delocalized electrons must be described using the space group of the crystal. The quasi-free electrons act as particles described by Bloch waves [4, 5] that interact with both the atoms in the lattice and with other quasi-free electrons. These exist in quantized states of energy termed electron energy bands [6, 7]. The highest filled band is called the valence band and the lowest unfilled band is the conduction band. The wave functions of the electrons in these bands must exhibit the periodic symmetry of the crystal lattice. This symmetry requirement results in the important concept of a forbidden band gap in the energy structure which is critical in determining whether the material is a conductor or not. In this chapter the atomic nuclei are assumed to be in a fixed periodic array. The effect of symmetry on the vibrational motion of these atoms was discussed in Chap. 7.

The concepts of cyclic groups are useful in treating wavelike excitations in solids such as electrons. The Schrödinger equation for a single electron described by Bloch function $\psi\left(\vec{r}\right)$ is

$$\left(-\frac{\hbar^2}{2m}\nabla^2 + V(\vec{r})\right)\psi\left(\vec{r}\right) = E\psi\left(\vec{r}\right). \tag{8.19}$$

The potential energy term is a periodic function of position that contains both a Coulomb and exchange part. This expression is invariant under translation operations of the symmetry group of the crystal lattice. Thus,

$$\vec{T}V(\vec{r}) = V(\vec{r}), \quad \vec{T}\psi(\vec{r}) = \sigma\psi(\vec{r}),$$

where σ is an eigenvalue for this wave function. Since the square of the wave function is the electron density,

$$\psi^*(\vec{r})\psi(\vec{r}) = \psi^*(\vec{r} - \vec{t})\psi(\vec{r} - \vec{t}) \text{ so } \sigma^*\sigma = 1.$$

Since the wave function in a specific direction is periodic over a dimension $L=Na$, where a is the lattice constant in that direction,

$$\vec{T}^n \psi(x) = \psi(x - na) = \sigma^n \psi(x)$$

and for $n=N$, the periodicity requires $\sigma^N=1$. One way to insure this condition is to define

$$\sigma = e^{2\pi i(n/N)}, \text{ where } n = 0, \pm 1, \pm 2, \ldots.$$

A wave vector can then be defined with the magnitude $k = (2\pi/a)(n/N)$. The eigenvalue then becomes $\sigma = e^{ik \cdot r}$. The solution to (8.19) is now written as

$$\psi\left(\vec{r}\right) = e^{ik \cdot \vec{r}} u\left(\vec{k}, r\right), \tag{8.20}$$

where $\vec{k}$ is the momentum of the electron in the crystal and $u\left(\vec{r}\right)$ is a function that contains the periodicity of the lattice

$$u\left(\vec{k}, \vec{r}\right) = u\left(\vec{k}, \vec{r} + \vec{t}\right), \tag{8.21}$$

where $\vec{t}$ is a lattice vector. This is called a *Bloch wave function* and it is subject to the condition $\psi\left(\vec{r}+\vec{t}\right) = e^{ik \cdot t}\psi\left(\vec{r}\right)$. Note the similarity between this eigenfunction for a quasiparticle wave in a crystal lattice and the basis function for a cyclic group discussed in Sect. 8.2.

A determinant of single electron Bloch functions can be formed to represent the multielectron wave function for the system (Hartree–Fock approximation). However, it can be shown that the symmetry properties of multielectron wave functions are the same as those of single electron wave functions [6] so treating only the latter is sufficient for the purposes of this book.

A symmetry element of the space group of the crystal is designated $\left\{\alpha|\vec{a}\right\}$ where α represents the rotation part of the operation and $\vec{a}$ represents the translation part. Using the fact that $\left\{\alpha|\vec{a}\right\}^{-1} = \left\{\alpha^{-1}|-\alpha^{-1}\vec{a}\right\}$, applying this operation to the Bloch wave function gives

$$\left\{\alpha|\vec{a}\right\}^{-1}\psi\left(\vec{k},\vec{r}\right) = e^{i\alpha k \cdot \left(\vec{r}-\vec{a}\right)} u(\alpha\vec{k},\vec{r}) = \psi\left(\alpha\vec{k},\vec{r}\right). \tag{8.22}$$

Thus this symmetry operation takes an electron wave function with wave vector $\vec{k}$ into a wave function with wave vector $\alpha\vec{k}$. Since this operation is caused by a symmetry operator for the system, its application does not change the energy, so $E(\alpha\vec{k}) = E(\vec{k})$. All of the points $\alpha\vec{k}$ that have the same energy eigenvalue and eigenvector are called [7] the *star of* $\vec{k}$. The number of vectors in the star can range between one and the order of the point group.

The different states of the system can be described in terms of the energies and wave functions described above by including a subscript $n=1,2,3,\dots$ to designate the states from the lowest to the highest energies. It is necessary to consider only the first Brillouin zone because of the periodicity of reciprocal space:

$$\psi_n\left(\vec{k}+\vec{K},\vec{r}\right) = \psi_n\left(\vec{k},\vec{r}\right) \tag{8.23}$$

and

$$E_n\left(\vec{k}+\vec{K}\right) = E_n\left(\vec{k}\right)$$

where $\vec{K}$ is a reciprocal lattice vector. Thus the electron wave functions are Bloch functions within the first Brillouin zone and periodic functions of $\vec{k}$ outside the first Brillouin zone.

In order to derive an expression for energy as a function of wave vector, it is assumed that the lattice potential acts as a weak perturbation on a free electron. Just as was the case of localized electronic states discussed in Chap. 4, this symmetric perturbation can lift the degeneracy of the free electron states. Under these conditions, the potential can be expressed as a Fourier series in the reciprocal lattice vector [3]

$$V\left(\vec{r}\right) = \sum_{\underline{K}} V_{\underline{K}} e^{i\vec{K}\cdot\vec{r}}. \tag{8.24}$$

This can be used in (8.19) with the wave function in (8.20) expressed as

$$\psi\left(\vec{k},\vec{r}\right) = \sum_{\underline{K}} u\left(\vec{K}\right) e^{i\left(\vec{k}+\vec{K}\right)\cdot\vec{r}}. \tag{8.25}$$

Equation (8.19) can then be solved for the coefficients $u\left(\vec{K}\right)$

$$\left[E - \left(\frac{\hbar^2}{2m}\right)\left(\vec{k}+\vec{K}\right)^2\right] u\left(\vec{K}\right) = \sum_{\underline{K}'} u\left(\vec{K}'\right) V_{\underline{K}-\underline{K}'}. \tag{8.26}$$

The solution for this can be written in two parts. For $\vec{K} = 0$,

$$E = \left(\frac{\hbar^2}{2m}\right)\vec{k}^2 + V_0 + \sum_{K\neq0} V_{-\underline{K}} \left[\frac{u\left(\vec{K}\right)}{u(0)}\right] \tag{8.27}$$

while for $\vec{K} \neq 0$

$$u\left(\vec{K}\right) = \frac{1}{E - \left(\frac{\hbar^2}{2m}\right)\left(\vec{k}+\vec{K}\right)^2} \sum_{\underline{K}'} u\left(\vec{K}'\right) V_{\underline{K}-\underline{K}'}. \tag{8.28}$$

In this weak perturbation approximation $E \gg V$ so $u\left(\vec{K}\right)\big/u(0) \ll 1$. Then only the dominant $\vec{K} = 0$ term in the sum in (8.28) is retained and the resulting expression for $u\left(\vec{K}\right)\big/u(0)$ substituted in (8.27). Approximating E in the denominator of (8.28) by the first term of (8.27), the final expression becomes

$$E \approx \left(\frac{\hbar^2}{2m}\right)k^2 + V_0 + \sum_{K\neq0} \frac{\left|V_{\underline{K}}\right|^2}{\left(\frac{\hbar^2}{2m}\right)\left[\vec{k}^2 - \left(\vec{k}+\vec{K}\right)^2\right]}. \tag{8.29}$$

Note that this has a singularity for specific reciprocal lattice vectors that obey the expression $\vec{k}^2 = \left(\vec{k}+\vec{K}\right)^2$ if $\left|V_{\underline{K}}\right|^2$ does not vanish. This occurs at the boundary of the Brillouin zone.

To elucidate the behavior of the energy band near the zone boundary, the secular determinant can be written as [5]

$$\begin{vmatrix} E_1 - E & V_{-K} \\ V_{-K} & E_2 - E \end{vmatrix} = 0. \tag{8.30}$$

Here E_1 is the energy at a wave vector $\vec{k} = \vec{k}_0 + \Delta\vec{k}$, where $\vec{k}_0$ is a general vector on the plane of the zone boundary and $\Delta\vec{k}$ is a small deviation from this plane. E_2 is the energy at wave vector $\vec{k} + \vec{K}$. From (8.29) these energies are

$$E_1 = E_b + \left(\frac{\hbar^2}{2m}\right)\left[\left(\Delta\vec{k}\right)^2 + 2\vec{k}_0 \cdot \Delta\vec{k}\right],$$

$$E_2 = E_b + \left(\frac{\hbar^2}{2m}\right)\left[\left(\Delta\vec{k}\right)^2 + 2\left(\vec{k}_0 + \vec{K}\right)\cdot \Delta\vec{k}\right], \tag{8.31}$$

where the unperturbed energy at the boundary is

$$E_b = \left(\frac{\hbar^2}{2m}\right)k_0^2 + V_0. \tag{8.32}$$

Expanding the determinant in (8.30) gives a quadratic equation for E that can be solved to give

$$E = E_b + \left(\frac{\hbar^2}{2m}\right)\left[\left(\Delta\vec{k}\right)^2 + \left(2\vec{k}_0 + \vec{K}\right)\cdot \Delta\vec{k}\right]$$
$$\pm \left\{\left[\left(\frac{\hbar^2}{2m}\right)\vec{K}\cdot\Delta\vec{k}\right]^2 + \left|V_{-K}\right|^2\right\}^{1/2}. \tag{8.33}$$

At the zone boundary $\Delta\vec{k} = 0$, so (8.33) simplifies to

$$E = E_b \pm \left|V_{-K}\right|. \tag{8.34}$$

This shows that the electron energy at the zone boundary has two values with a gap of $2\left|V_{-K}\right|$.

Figure 8.7 shows schematically the variation of energy with wave vector described by (8.29). Near the center of the Brillouin zone the energy varies quadratically with k while at the zone boundary it flattens out and becomes double valued with an energy gap. Since the velocity of the electron is proportional to the slope of

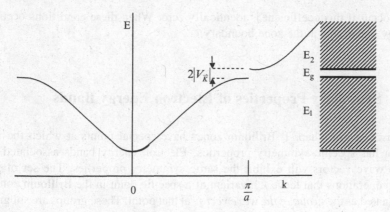

Fig. 8.7 Energy versus wave vector

the dispersion curve, the electron approaches zero velocity at the zone boundary. Projecting these dispersion curves onto the energy axis shows the allowed energy states of the electron in the lattice. The density of states is large so the discrete states become bands of energies. The double-valued energy at the Brillouin zone boundary produces a forbidden energy band or band gap. The lowest energy band completely filled with electrons is the valance band while the lowest partially filled band is the conduction band. The band gap plays an important role in determining the electrical conductivity properties of the material.

The symmetry operations of the point group that operate on a position vector in real space are the same as those that operate on the wave vector in reciprocal space. A specific operation on $\vec{k}$ in reciprocal space produces the same effect as its inverse operation produces on $\vec{r}$ in real space. The translation operations on the position vector in real space do not have an equivalent operation on the wave vector in reciprocal space. The symmetry properties of the lattice require that a space group operation $\left\{\alpha|\vec{\tau}\right\}$ operating on the lattice potential $V\left(\vec{r}\right)$ gives

$$\left\{\alpha|\vec{\tau}\right\}V\left(\vec{r}\right) = V(\alpha\vec{r}+\vec{\tau}). \tag{8.35}$$

Using (8.18), this becomes

$$\left\{\alpha|\vec{\tau}\right\}V_{\vec{K}} = e^{i\alpha\vec{K}\cdot\vec{\tau}}V_{\varepsilon\vec{K}}. \tag{8.36}$$

For a symmorphic space group the operators $\{\alpha|0\}$ form a subgroup. In this case, $V_{\vec{K}} = V_{\varepsilon\vec{K}}$ so all coefficients are nonzero. For a nonsymmorphic space group with an operation for which $\alpha\vec{K} = \vec{K}$ and $\exp(i\vec{K}\cdot\vec{\tau}) = -1$, (8.36) shows that the results of applying this symmetry operation to the lattice potential gives $V_{\vec{K}} = -V_{\vec{K}}$. This can only occur if this coefficient is identically zero. When these conditions occur, no energy gap occurs at the zone boundary.

8.4   Symmetry Properties of Electron Energy Bands

As discussed in Chap. 1, Brillouin zones have special points at which the wave vector has specific symmetry properties. Electron energy bands associated with these wave vectors will exhibit the same symmetry properties. The set of point group operations that leave $\bar{k}$ invariant at a specific point in the Brillouin zone are designated as the *group of the wave vector* at that point. These groups are subgroups of the total point symmetry group of the crystal. Applying these operations to $\bar{k}$ form a set of wave vectors that is called the *star of* $\bar{k}$.

For example, Fig. 1.10 shows the Brillouin zone for a simple cubic lattice with the points of special symmetry of $\bar{k}$ designated. Consider a wave vector pointed along the k_x axis. This will have the special symmetry designated by Δ which was shown to be C_{4v}. The symmetry elements of this point group are shown in Table 2.11 to be E, $2C_4$, C_2, $2\sigma_v$, and $2\sigma_d$. Figure 8.8 shows the star of the wave vector at the Δ point in the first Brillouin zone of simple cubic symmetry.

Energy bands are characterized by their $\bar{k}$ vector in the first Brillouin zone. The star of $\bar{k}$ defines a set of basis functions. Going along a direction of high symmetry in the Brillouin zone, the group of $\bar{k}$ changes to different subgroups at different points. Compatibility relations allow the connection of different bands at different points. For example, Fig. 8.9 shows a schematic diagram of moving from the center of a Brillouin zone of a crystal with cubic symmetric along the Δ direction to the X point on the surface as shown in Fig. 1.6.

Several different theoretical approaches have been developed to derive mathematical expressions for electron energy bands [5, 7]. The most sophisticated ones involve computer simulation techniques. One simple approach that demonstrates the effects of symmetry is to use a local coordinate system at the

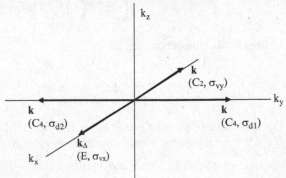

Fig. 8.8 Star of the wave vector at the Δ point in the first Brillouin zone with simple cubic symmetry

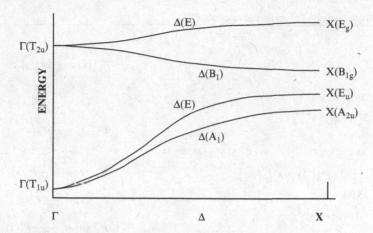

Fig. 8.9 Schematic representation of electron energy bands in one part of the Brillouin zone of a crystal with octahedral symmetry

site of each atom to determine the charge density and crystal potential. This has to reflect the point group symmetry of the crystal at each atom. The symmetrized crystalline potential then acts on each electron to give the energy bands. In this approach the most general expression for the electron energy can be written as [2]

$$E\left(\vec{k}\right) = \sum_i^3 \frac{\hbar^2 k_i^2}{2m_{ii}^*} + \sum_{n=0}^\infty \sum_{m=-n}^n C_{nm} k^n Y_n^m(\theta, \varphi). \tag{8.37}$$

The second term is the crystalline potential expressed as a multipole expansion in terms of spherical harmonic functions. This is the standard way to express the Coulomb interaction between two charged particles, in this case an electron and a lattice ion [9]. The spherical harmonic functions $Y_n^m(\theta, \varphi)$ can be written as a product of a normalization factor, an associated Legendre polynomial $P_n^m(\cos \theta)$, and exponential function

$$Y_n^m(\theta, \varphi) = N_{nm}^{-1/2} P_n^m(\cos \theta) e^{im\varphi}.$$

The first several associated Legendre polynomials are shown in Table 8.6.

The crystalline potential must be real so the product of the real and imaginary parts of the exponential function and the complex factor C_{nm} can be separated to give

$$E\left(\vec{k}\right) = \sum_i^3 \frac{\hbar^2 k_i^2}{2m_{ii}^*} + \sum_{n=0}^\infty \sum_{m=-n}^n k^n P_n^m(\cos \theta) \left[C_{nm}^{re} \cos m\varphi + C_{nm}^{im} \sin m\varphi\right]. \tag{8.38}$$

The first term in (8.32) represents the contribution from kinetic energy. In a crystal the effective mass of an electron can be different in different directions so this is written as the diagonal components of a tensor [3, 5, 6].

Table 8.6 Associated Legendre polynomials $P_n^m(x)$

$$P_0^0(x) = 1$$
$$P_1^0(x) = x$$

$$P_1^1(x) = -(1-x^2)^{1/2}$$ $$P_1^{-1}(x) = -\tfrac{1}{2}P_1^1(x)$$

$$P_2^0(x) = \tfrac{1}{2}(3x^2 - 1)$$

$$P_2^1(x) = -3x(1-x^2)^{1/2}$$ $$P_2^{-1}(x) = -\frac{1}{6}P_2^1(x)$$
$$P_2^2(x) = 3(1-x^2)$$
$$P_2^{-2}(x) = \frac{1}{24}P_2^2(x)$$

$$P_3^0(x) = \tfrac{1}{2}(5x^3 - 3x)$$

$$P_3^1(x) = -\frac{3}{2}(5x^2 - 1)(1-x^2)^{1/2}$$ $$P_3^{-1}(x) = -\frac{1}{12}P_3^1(x)$$
$$P_3^2(x) = 15x(1-x^2)$$ $$P_3^{-2}(x) = \frac{1}{120}P_3^2(x)$$
$$P_3^3(x) = -15(1-x^2)^{3/2}$$ $$P_3^{-3}(x) = -\frac{1}{720}P_3^3(x)$$

$$P_4^0(x) = \tfrac{1}{8}(35x^4 - 30x^2 + 3)$$

$$P_4^1(x) = -\frac{5}{2}(7x^3 - 3x)(1-x^2)^{1/2}$$ $$P_4^{-1}(x) = -\frac{1}{20}P_4^1(x)$$

$$P_4^2(x) = \frac{15}{2}(7x^2 - 1)(1-x^2)$$ $$P_4^{-2}(x) = \frac{1}{360}P_4^2(x)$$

$$P_4^3(x) = -105x(1-x^2)^{3/2}$$ $$P_4^{-3}(x) = -\frac{1}{5040}P_4^3(x)$$
$$P_4^4(x) = 105(1-x^2)^2$$
$$P_4^{-4}(x) = \frac{1}{40320}P_4^4(x)$$

There are two additional restrictions on the energy $E\left(\vec{k}\right)$ that determine what terms appear in the expansion of the crystalline potential in (8.38). First it must be invariant under time reversal, so $E\left(-\vec{k}\right) = E\left(\vec{k}\right)$. Due to the factor of k^n, this requires that only even values of n ($n=0,2,4,\ldots$) appear in the expansion. In addition, only the even parity associated Legendre polynomials can be present so m must also be even. The second requirement is that $E\left(\vec{k}\right)$ is invariant with respect to all of the symmetry elements that make up the group of the $\vec{k}$ vector at a specific point in reciprocal space.

As an example, consider a point in reciprocal space at which $\vec{k}$ has C_{2v} symmetry. The energy band has the form [8]

$$E\left(\vec{k}\right) = \frac{\hbar^2}{2}\left[\frac{k_x^2}{m_{xx}^*} + \frac{k_y^2}{m_{yy}^*} + \frac{k_z^2}{m_{zz}^*}\right] + C_{20}^{\text{re}}k^2 P_2^0(\cos\theta) + C_{22}^{\text{re}}k^2 P_2^2(\cos\theta)\cos 2\varphi$$

$$+ C_{40}^{\text{re}}k^4 P_4^0(\cos\theta) + C_{42}^{\text{re}}k^4 P_4^2(\cos\theta)\cos 2\varphi + C_{44}^{\text{re}}k^4 P_4^4(\cos\theta)\cos 4\varphi + \cdots.$$

$$(8.39)$$

For C_{2v} symmetry there are three distinct directions for the effective mass. All of the associated Legendre polynomials with even values of n and m appear in the expansion. The term with $n=0$, $m=0$ shifts the energy without any dependence on the angles θ and φ so it is neglected in the expansion. The C_{2v} symmetry group has a twofold rotation about the z-axis and reflections in the xz and yz planes. All three of these symmetry elements leave θ unchanged so they do not restrict the $P_n^m(\cos\theta)$ appearing in the expansion. These symmetry elements also leave $\cos(m\varphi)$ unchanged. However, the mirror planes change $\sin(m\varphi)$ into $-\sin(m\varphi)$ so the coefficients of terms in the expansion with this factor must be zero.

As another example, consider a point in reciprocal space at which $\bar{k}$ has D_{2d} symmetry. In this case the energy band has the form

$$
\begin{aligned}
E\left(\bar{k}\right) = \frac{\hbar^2}{2}\left[\frac{k_{//}^2}{m_{//}^*} + \frac{k_\perp^2}{m_\perp^*}\right] + C_{20}^{re}k^2 P_2^0(\cos\theta) \\
+ C_{40}^{re}k^4 P_4^0(\cos\theta) + C_{44}^{re}k^4 P_4^4(\cos\theta)\cos 4\varphi + \cdots.
\end{aligned}
\tag{8.40}
$$

For D_{2d} symmetry there are only two directions for the effective mass, parallel and perpendicular to the major axis of rotation. The rest of the expansion is similar to the expression in (8.39) except that the terms for $n=2$, $m=2$ and $n=4$, $m=2$ are missing. The symmetry elements for D_{2d} are a major twofold axis of rotation, two other C'_2 axes perpendicular to this, a mirror plane bisecting these two axes and containing the major rotation axis, and an S_4 rotation–reflection operation. This latter symmetry element takes φ into $(\varphi+\pi/2)$ so the sine and cosine functions for 2φ both transform into minus themselves and thus must have zero expansion coefficients.

A full treatment of the many-body problem of electron energy band theory requires a choice of the form of the wave functions such as LCAO (linear combination of atomic orbitals) or LAPW (linear augmented plane waves). These calculations are beyond the scope of this book and are the subject of complete text books such as [7] by Slater. The interested reader is referred to these books for further information. The effects of symmetry on wavelike particles in crystals can be applied in a similar way to entities other than electrons. Chapter 7 describes these principles as applied to lattice vibration phonons.

8.5 Problems

1. Consider a crystal with a simple cubic lattice. Determine the group of the k vector and draw the star of the k vector at the X point at the center of the face of the Brillouin zone.
2. Using the "tight binding approximation," the eigenfunctions given in (8.20) are expressed in terms of atomic orbitals as $\psi_{Ak} = N^{-1/2}\sum_j e^{ik_x x_j}\varphi_A\left(\bar{r} - \bar{R}_j\right)$ where the last factor in this expression is an atomic orbital. For the crystal described in problem 1, write the eigenfunctions for atomic s and p orbitals at the Δ point of the Brillouin zone. Show how these functions transform in the group of the k

vector at this point and draw the energy bands going from the Γ point along the Δ direction to the X point.

3. In the "free electron approximation," the potential $V(\mathbf{r})$ in (8.19) is taken to be equal to zero everywhere. The solutions of (8.19) are then eigenfunctions and eigenvalues of the form

$$\psi_{\vec{k}l} = e^{i\left(\vec{k}-\vec{K}_l\right)\cdot\vec{r}}, \quad E_l\left(\vec{k}\right) = \frac{\hbar^2}{2m}\left|\vec{k}-\vec{K}_l\right|^2.$$

In these expressions $\vec{k} = (2\pi/a)(\alpha,\beta,\gamma)$ where α, β, and γ are numbers greater than 0 and less than 1, and $\vec{K} = (2\pi/a)(L_1,L_2,L_3)$ where the L_i are integers. For a face-centered cubic crystal structure $L_1=(-l_1+l_2+l_3)$, $L_2=(l_1-l_2+l_3)$, and $L_3=(l_1+l_2-l_3)$ where the values of the l_i are integers. Find a figure in the literature of the first Brillouin zone for a face-centered cubic crystal and write the k vectors, eigenfunctions, and eigenvalues in terms of the α, β, γ, and L_i parameters at the Γ, Δ, and X points of the Brillouin zone.

4. Determine the L vectors for the crystal in problem 3 that give the five lowest energy eigenvalues at the Γ and X points in the Brillouin zone. What are the energies and degeneracies of each of these electron band states.

5. Determine the common L vectors for the Γ and X points in the Brillouin zone in problem 4 and use these L vectors in the expression for the eigenfunction at the Δ point in the Brillouin zone found in problem 3 to derive an expression for the electron energy bands going from the Γ point to the X point along the Δ direction. Draw a diagram to scale showing these energy bands.

6. Write the set of four eigenfunctions for the energy band $E(\Delta) = 2ma^2/\hbar^2 = (\alpha - 1)^2 + 2$ found in problem 5. Show how these functions transform under the operations of the group of the wave vector at the Δ point in the Brillouin zone. Express these eigenfunctions in terms of the irreducible representations of the group.

References

1. F.A. Cotton, *Chemical Applications of Group Theory* (Wiley, New York, 1963)
2. C.J. Ballhausen, *Introduction to Ligand Field Theory* (McGraw-Hill, New York, 1962)
3. R.S. Knox, A. Gold, *Symmetry in the Solid State* (Benjamin, New York, 1964)
4. C. Kittel, *Solid State Physics* (Wiley, New York, 1976)
5. W.A. Harrison, *Solid State Theory* (McGraw Hill, New York, 1970)
6. M. Lax, *Symmetry Principles in Solid State and Molecular Physics* (Wiley, New York, 1974)
7. J.C. Slater, *Symmetry and Energy Bands in Crystals* (Dover, New York, 1972)
8. M. Sachs, *Solid State Theory* (McGraw, New York, 1963)
9. J.D. Jackson, *Classical Electrodynamics* (Wiley, New York, 1975)

Chapter 3
Tensor Properties of Crystals

R.C. Powell, *Symmetry, Group Theory, and the Physical Properties of Crystals*,
Lecture Notes in Physics 824, DOI 10.1007/978-1-4419-7598-0, pp. 55–78,
© Springer Science+Business Media, LLC 2010

DOI 10.1007/978-1-4419-7598-0_9

On page 72, Table 3.5 has the forms of 3rd rank tensors for all the crystallographic point groups. Five of these have factors of 2 in some of the components. These factors of 2 can be found in other published tables of 3rd rank tensors but they are only valid for a different type of notation that is not used in the book. To be consistent with the tensor notation used throughout this book, no factors of 2 should appear in these tensors. The Erratum provides a corrected Table 3.5 with the factors of 2 eliminated from all the tensor elements.

In addition, on page 73 an example is given in the second paragraph using one of the tensors from Table 3.5 that contains erroneous factors of 2 in some of its components. The Erratum provides a corrected paragraph without the factors of 2.

The online version of the original chapter can be found at
http://dx.doi.org/10.1007/978-1-4419-7598-0_3

Table 3.5 Form of third rank tensors for the crystallographic point groups

$$C_i, C_{2h}, D_{2h},$$
$$C_{4h}, D_{4h}, S_6,$$
$$D_{3d}, C_{6h}, D_{6h},$$

C_1

$$\begin{bmatrix} d_{111} & d_{121} & d_{131} \\ d_{112} & d_{122} & d_{132} \\ d_{113} & d_{123} & d_{133} \\ d_{211} & d_{221} & d_{231} \\ d_{212} & d_{222} & d_{232} \\ d_{213} & d_{223} & d_{233} \\ d_{311} & d_{321} & d_{331} \\ d_{312} & d_{322} & d_{332} \\ d_{313} & d_{323} & d_{333} \end{bmatrix}$$

T_h, O, O_h

$$\begin{bmatrix} 0 & 0 & 0 \\ 0 & 0 & 0 \\ 0 & 0 & 0 \\ 0 & 0 & 0 \\ 0 & 0 & 0 \\ 0 & 0 & 0 \\ 0 & 0 & 0 \\ 0 & 0 & 0 \\ 0 & 0 & 0 \end{bmatrix}$$

C_2

$$\begin{bmatrix} 0 & d_{121} & 0 \\ d_{121} & 0 & d_{132} \\ 0 & d_{132} & 0 \\ d_{211} & 0 & d_{231} \\ 0 & d_{222} & 0 \\ d_{231} & 0 & d_{233} \\ 0 & d_{321} & 0 \\ d_{321} & 0 & d_{332} \\ 0 & d_{332} & 0 \end{bmatrix}$$

C_{2v}

$$\begin{bmatrix} 0 & 0 & d_{131} \\ 0 & 0 & 0 \\ d_{131} & 0 & 0 \\ 0 & 0 & 0 \\ 0 & 0 & d_{232} \\ 0 & d_{232} & 0 \\ d_{311} & 0 & 0 \\ 0 & d_{322} & 0 \\ 0 & 0 & d_{333} \end{bmatrix}$$

C_s

$$\begin{bmatrix} d_{111} & 0 & d_{131} \\ 0 & d_{122} & 0 \\ d_{131} & 0 & d_{133} \\ 0 & d_{221} & 0 \\ d_{221} & 0 & d_{232} \\ 0 & d_{232} & 0 \\ d_{311} & 0 & d_{331} \\ 0 & d_{322} & 0 \\ d_{331} & 0 & d_{333} \end{bmatrix}$$

D_2

$$\begin{bmatrix} 0 & 0 & 0 \\ 0 & 0 & d_{132} \\ 0 & d_{132} & 0 \\ 0 & 0 & d_{231} \\ 0 & 0 & 0 \\ d_{231} & 0 & 0 \\ 0 & d_{321} & 0 \\ d_{321} & 0 & 0 \\ 0 & 0 & 0 \end{bmatrix}$$

C_4

$$\begin{bmatrix} 0 & 0 & d_{131} \\ 0 & 0 & d_{132} \\ d_{131} & d_{132} & 0 \\ 0 & 0 & -d_{132} \\ 0 & 0 & d_{131} \\ -d_{132} & d_{131} & 0 \\ d_{311} & 0 & 0 \\ 0 & d_{311} & 0 \\ 0 & 0 & d_{333} \end{bmatrix}$$

D_4

$$\begin{bmatrix} 0 & 0 & 0 \\ 0 & 0 & d_{132} \\ 0 & d_{132} & 0 \\ 0 & 0 & d_{231} \\ 0 & 0 & 0 \\ d_{231} & 0 & 0 \\ 0 & 0 & 0 \\ 0 & 0 & 0 \\ 0 & 0 & 0 \end{bmatrix}$$

S_4

$$\begin{bmatrix} 0 & 0 & d_{131} \\ 0 & 0 & d_{132} \\ d_{131} & d_{132} & 0 \\ -d_{131} & 0 & d_{132} \\ 0 & 0 & 0 \\ d_{132} & 0 & 0 \\ d_{311} & d_{321} & 0 \\ d_{321} & -d_{311} & 0 \\ 0 & 0 & 0 \end{bmatrix}$$

C_{4v}

$$\begin{bmatrix} 0 & 0 & d_{131} \\ 0 & 0 & 0 \\ d_{131} & 0 & 0 \\ 0 & 0 & 0 \\ 0 & 0 & d_{131} \\ 0 & d_{131} & 0 \\ d_{311} & 0 & 0 \\ 0 & d_{311} & 0 \\ 0 & 0 & d_{333} \end{bmatrix}$$

D_{2d}

$$\begin{bmatrix} 0 & 0 & 0 \\ 0 & 0 & d_{132} \\ 0 & d_{132} & 0 \\ 0 & 0 & d_{132} \\ 0 & 0 & 0 \\ d_{132} & 0 & 0 \\ 0 & d_{321} & 0 \\ d_{321} & 0 & 0 \\ 0 & 0 & 0 \end{bmatrix}$$

C_3

$$\begin{bmatrix} d_{111} & d_{211} & d_{131} \\ d_{211} & -d_{111} & d_{132} \\ d_{131} & d_{132} & 0 \\ d_{211} & -d_{111} & -d_{132} \\ -d_{111} & -d_{211} & d_{131} \\ -d_{132} & d_{131} & 0 \\ d_{311} & 0 & 0 \\ 0 & d_{311} & 0 \\ 0 & 0 & d_{333} \end{bmatrix}$$

D_3

$$\begin{bmatrix} d_{111} & 0 & 0 \\ 0 & -d_{111} & d_{132} \\ 0 & d_{132} & 0 \\ 0 & -d_{111} & -d_{132} \\ -d_{111} & 0 & 0 \\ -d_{132} & 0 & 0 \\ 0 & 0 & 0 \\ 0 & 0 & 0 \\ 0 & 0 & 0 \end{bmatrix}$$

C_{3v}

$$\begin{bmatrix} 0 & -d_{222} & d_{131} \\ -d_{222} & 0 & 0 \\ d_{131} & 0 & 0 \\ -d_{222} & 0 & 0 \\ 0 & d_{222} & d_{131} \\ 0 & d_{131} & 0 \\ d_{311} & 0 & 0 \\ 0 & d_{311} & 0 \\ 0 & 0 & d_{333} \end{bmatrix}$$

C_{3h}

$$\begin{bmatrix} d_{111} & -d_{222} & 0 \\ -d_{222} & -d_{111} & 0 \\ 0 & 0 & 0 \\ -d_{222} & d_{111} & 0 \\ d_{111} & d_{222} & 0 \\ 0 & 0 & 0 \\ 0 & 0 & 0 \\ 0 & 0 & 0 \\ 0 & 0 & 0 \end{bmatrix}$$

D_{3h}

$$\begin{bmatrix} 0 & -d_{222} & 0 \\ -d_{222} & 0 & 0 \\ 0 & 0 & 0 \\ -d_{222} & 0 & 0 \\ 0 & d_{222} & 0 \\ 0 & 0 & 0 \\ 0 & 0 & 0 \\ 0 & 0 & 0 \\ 0 & 0 & 0 \end{bmatrix}$$

C_6

$$\begin{bmatrix} 0 & 0 & d_{131} \\ 0 & 0 & d_{132} \\ d_{131} & d_{132} & 0 \\ 0 & 0 & -d_{132} \\ 0 & 0 & d_{131} \\ -d_{132} & d_{131} & 0 \\ d_{311} & 0 & 0 \\ 0 & d_{311} & 0 \\ 0 & 0 & d_{333} \end{bmatrix}$$

C_{6v}

$$\begin{bmatrix} 0 & 0 & d_{232} \\ 0 & 0 & 0 \\ d_{232} & 0 & 0 \\ 0 & 0 & 0 \\ 0 & 0 & d_{232} \\ 0 & d_{232} & 0 \\ d_{311} & 0 & 0 \\ 0 & d_{311} & 0 \\ 0 & 0 & d_{333} \end{bmatrix}$$

D_6

$$\begin{bmatrix} 0 & 0 & 0 \\ 0 & 0 & d_{132} \\ 0 & d_{132} & 0 \\ 0 & 0 & -d_{132} \\ 0 & 0 & 0 \\ -d_{132} & 0 & 0 \\ 0 & 0 & 0 \\ 0 & 0 & 0 \\ 0 & 0 & 0 \end{bmatrix}$$

T, T_d

$$\begin{bmatrix} 0 & 0 & 0 \\ 0 & 0 & d_{132} \\ 0 & d_{132} & 0 \\ 0 & 0 & d_{132} \\ 0 & 0 & 0 \\ d_{132} & 0 & 0 \\ d_{132} & 0 & 0 \\ 0 & d_{132} & 0 \\ d_{132} & 0 & 0 \end{bmatrix}$$

Page 73, second paragraph

As a practical example, consider a quartz crystal that has D_3 symmetry at room temperature. The piezoelectric effect for this case is given by

$$
\begin{pmatrix} P_1 \\ P_2 \\ P_3 \end{pmatrix} = \begin{bmatrix} d_{111} & 0 & 0 \\ 0 & -d_{111} & d_{132} \\ 0 & d_{132} & 0 \\ 0 & -d_{111} & -d_{132} \\ -d_{111} & 0 & 0 \\ -d_{132} & 0 & 0 \\ 0 & 0 & 0 \\ 0 & 0 & 0 \\ 0 & 0 & 0 \end{bmatrix} \begin{pmatrix} \sigma_{11} & \sigma_{12} & \sigma_{13} \\ \sigma_{21} & \sigma_{22} & \sigma_{23} \\ \sigma_{31} & \sigma_{32} & \sigma_{33} \end{pmatrix},
$$

so

$$
P_1 = d_{111}\sigma_{11} - d_{111}\sigma_{22} + d_{132}\sigma_{32} + d_{132}\sigma_{23} = (\sigma_{11} - \sigma_{22})d_{111} + (\sigma_{32} + \sigma_{23})d_{132}
$$
$$
P_2 = -d_{111}\sigma_{21} - d_{132}\sigma_{31} - d_{111}\sigma_{12} - d_{132}\sigma_{13} = -d_{111}(\sigma_{21} + \sigma_{12}) - (\sigma_{13} + \sigma_{31})d_{132}
$$
$$
P_3 = 0
$$

If a uniaxial stress is applied in the σ_{11} direction, $P_1 = d_{111}\sigma_{11}$ and $P_2 = 0$. The same tensile stress applied along σ_{22} also produces a polarization along P_1. The two-fold rotation axis P_1 is the electric axis of quartz. Shear stress can produce polarization along P_2 but no stress conditions can produce a polarization along P_3.

Index

A
Axial vector, crystallographic point groups,
61–62

B
Basis functions, 31
Birefringence, 114–118
Bloch wave function, 216–218
Bravais lattices, 6
Brillouin zones, 21–23

C
Cartesian coordinates transformation, 28–29
Character table
    32 crystallographic point groups, 32–38
    Pauli spin operators, 39–40
Coordinates transformation matrix, 29
Crystal field symmetry
    medium crystal field case, 86
    octahedral coordination, ligands, 86–87
    perturbation theory, 86
    Stark splitting, 85
    strong crystal field case, 86
    weak crystal field case, 85
Crystals optical properties and symmetry
    birefringence
        crystal dielectric property, 114–115
        double refraction, 115
        indicatrix, 116–117
        linear optics fundamental equation,
            115–116
        optically anisotropic material
            properties, 118
        uniaxial crystal, wave surface, 118
    electrooptical effect
        crystallographic point groups,
            electrooptic tensor, 126

crystallographic point groups, Kerr
    tensor, 130–131
crystals electrooptic coefficients, 127
C_{3v} symmetry crystal electric field, 129
D_{2d} symmetry electrooptic tensor, 128
first-order electrooptic effect/Pockels
    effect, 123–124
Kerr coefficients, 129
Kerr effect, 124–125
optical activity
    definition, 118
    gyration tensor, 120
    gyration tensor crystallographic point
        groups, 122–123
    Neumann's principle, 122
    rotatory dispersion effect, 121
    rotatory power, 119
    spatial dispersion, 119
    symmetry properties, gyration tensor
        components, 122
photoelastic effect
    elastooptical coefficients, 134
    photoelastic tensor-tensor form
        difference, 133
    piezooptical coefficients, 131
    refractive index coefficients, 132
polarization, tensor treatment
    electromagnetic light wave, electric
        field, 106
    Jones matrices, 112
    Jones vector, 110–111
    left-circularly polarized light, 107
    light wave electric field, 106–107
    Mueller matrices, 113, 114
    polarization transformation matrices,
        113
    polarization vectors, 109

R.C. Powell, *Symmetry, Group Theory, and the Physical Properties of Crystals*,
Lecture Notes in Physics 824, DOI 10.1007/978-1-4419-7598-0,
© Springer Science+Business Media, LLC 2010